# STRUCTURES OF DISCRETE EVENT SIMULATION:
## An Introduction to the Engagement Strategy

# ELLIS HORWOOD BOOKS IN COMPUTING SCIENCE

*General Editors:* Professor JOHN CAMPBELL, University College London, and
BRIAN L. MEEK, King's College London (KQC), University of London

## Ellis Horwood Series in Artificial Intelligence

*Series Editor:* Professor JOHN CAMPBELL, Department of Computer Science,
University College London

*\* In preparation*

# STRUCTURES OF DISCRETE EVENT SIMULATION:
## An Introduction to the Engagement Strategy

J. B. EVANS, Ph.D.
Lecturer in Computer Science
Department of Computer Science
University of Hong Kong

**ELLIS HORWOOD LIMITED**
Publishers · Chichester

Halsted Press: a division of
**JOHN WILEY & SONS**
New York · Chichester · Brisbane · Toronto

First published in 1988 by
**ELLIS HORWOOD LIMITED**
Market Cross House, Cooper Street,
Chichester, West Sussex, PO19 1EB, England
*The publisher's colophon is reproduced from James Gillison's drawing of the ancient Market Cross, Chichester.*


**Distributors:**

*Australia and New Zealand:*
JACARANDA WILEY LIMITED
GPO Box 859, Brisbane, Queensland 4001, Australia

*Canada:*
JOHN WILEY & SONS CANADA LIMITED
22 Worcester Road, Rexdale, Ontario, Canada

*Europe and Africa:*
JOHN WILEY & SONS LIMITED
Baffins Lane, Chichester, West Sussex, England

*North and South America and the rest of the world:*
Halsted Press: a division of
JOHN WILEY & SONS
605 Third Avenue, New York, NY 10158, USA

*South-East Asia*
JOHN WILEY & SONS (SEA) PTE LIMITED
37 Jalan Pemimpin # 05–04
Block B, Union Industrial Building, Singapore 2057

*Indian Subcontinent*
WILEY EASTERN LIMITED
4835/24 Ansari Road
Daryaganj, New Delhi 110002, India

© **1988 J.B. Evans/Ellis Horwood Limited**

**British Library Cataloguing in Publication Data**
Evans, J. B. (John B.), *1947–*
Structures of discrete event siluation: an introduction to the engagement strategy. —
(Ellis Horwood series in artificial intelligence).
1. Discrete events. Mathematical models.
Applications of computer systems
I. Title
003

**Library of Congress Card No.** 88–6826

ISBN 0–7458–0103–X (Ellis Horwood Limited)
ISBN 0–470–21097–4 (Halsted Press)

Phototypeset in Times by Ellis Horwood Limited
Printed in Great Britain by Hartnolls, Bodmin

# Table of contents

# Preface

Compared to the majority of books available on simulation, we stress an alternative perspective. As the title of this book indicates, we are concerned with the *structural* aspects of discrete-event simulation, by which we tackle the essentially *linguistic* problem of describing a dynamic system in formal terms which can be used to define a computer program. Although discrete-event simulation emerged out of a basically statistical model, its techniques are now applied to areas far from where steady-state arguments are of any practical use. Consequently, we can understand the practice of simulation as a component of computing practice, through the design of software, algorithms, methodologies and languages.

Being one of the earliest applications of computers to management and industrial problems, simulation has frequently proved innovative in the field of software. In the early 1960s the simulation languages GPSS (Gordon, 1978) and SIMSCRIPT (Markowitz *et al.*, 1963) were among the first attempts to produce high-level languages, GPSS seeking to retain a close relationship with diagrams, and SIMSCRIPT emulating natural language. Knuth (1973b) recalls that SIMSCRIPT gave rise to the buddy system for dynamic storage allocation. The requirement to service future-event notices in a priority queue has been innovatory in the development of the *heap* (Williams, 1963, 1964) and, recently, other newer heap-like structures (Jones, 1986).

The practice of simulation popularised the concept of a *coroutine* (Conway, 1963), or *process,* primarily through its adoption by Simula (Nygaard and Dahl, 1978), which has proved especially valuable in the field of operating systems. Simula also made early advances in the area of abstract data types, with its adoption of what has come to be known as the object-oriented programming style.

Despite its innovatory character, which derives in part from the breadth of its application area, simulation still has in many people's ears a rather old-fashioned ring. Perhaps this is due to its reputation as a 'last resort' only to be undertaken when more analytic procedures turn out to be intractable. In fact, simulation may give understanding and insight in situations where statistics cannot be applied. As a general 'what-if' hypothesis tester, simulation has an important role to play as an integrated element in future computer systems.

Therefore we seek to situate the study of simulation more completely within the Computer Science discipline. Its need for programs extending further than just algorithms, its relationship with methodology, and just its breadth of application make simulation intrinsically interesting for students. It is challenging for the language designer, requiring a more general approach to the study of parallel processing; in simulation complexity, we are skimming the surface of a phenomenology of parallelism. Encouraging criticism of current methods and languages, some of which have a long pedigree, can also bring out a fresh attitude to the acceptance of system software.

As a counterbalance to the approach of this book, the reader might like to read the predominantly statistical works of Yakowitz (1977) and Bratley *et al.* (1983); other books which tend to concentrate their attention more on structural aspects are Birtwistle (1979) and Kreutzer (1986).

We start in Chapter 1 with a look at the basic relationship between simulation and modelling as a whole. We view simulation as the activity of mapping a real dynamic onto the control flow of a computer program. In Chapter 2, different ways of representing the dynamic are discussed, as we narrow in on our main field, that of discrete-event simulation, defined with respect to the two paradigms. Chapter 3 deals with the algorithmic problems of discrete-event time advance. In Chapter 4, we compare the three main strategies for representing, in program form, the unfolding of the dynamic.

Methods of simulating randomness are taken up in Chapter 5. The linguistic approach to dynamic representation, as expressed in a variety of languages, is pursued in Chapter 6, with a characterisation of types of simulation complexity in Chapter 7. In Chapter 8 a new strategy, the engagement strategy, is introduced along with the SIMIAN language, whose implementation is supported by wait-until semantics. Finally in Chapter 9 we look at future developments in simulation, especially as it interacts and integrates with new ideas from artificial intelligence and novel programming paradigms.

This book will be of general interest to all concerned with computing, and its application; knowledge of a major high-level language, such as Pascal, is assumed. It will be found useful to practising simulationists, for whom it may serve as a window through which to examine the interior of the software whose validity they may take for granted. It has been derived largely from lecture notes for undergraduates in computer science at the second and third years, although many ideas go back to my Research Associateship at the Centre in Simulation at the University of Lancaster, 1978–1981.

I am grateful to those who have commented on earlier drafts, amongst whom I would like to specifically thank John Bacon-Shone, D. W. Jones, Ernie Jordan, J. H. Kingston, Dick Nance, C. M. Reeves, Pat Rivett, and John Campbell, the editor of the series. Naturally, the responsibility for inaccuracies and obscurities lies with me, the author. I would also like to extend my thanks to present and past simulation students of COMP405, CS60 and CS261, who bore the brunt of these ideas; my project and research students: Victor Lam, Wong Lap-shun, Sylvia Chan, Daniel Au Yeung and David Tam, who helped me to develop them, and especially to my wife, Sue, who encouraged me to write them down.

*Hong Kong,*
*December 1987*                                              John B. Evans

# 1

# Models and computation

## 1.1 BASICS OF MODELLING

It may be said that Man differentiates himself from the animals insofar as he exerts control over his natural surroundings creating, to some extent, his own environment. The means by which this control is exercised has led to a continual process of technological development, starting from primitive artefacts such as stone tools and spears, up through the harnessing of natural forces, to the present-day computer. Along with the technology has come the simultaneous development of the intellect.

Palaeolithic art, such as representations of hunting scenes, can be interpreted as an expression of the wish to control Nature, providing the technological advances with a purpose around which to crystallise and develop. The projection of external reality, or perhaps some desirable version of it, in the form of a computer program can be seen as a part of twentieth-century Man's need to influence his surroundings, and the computer modelling of dynamic reality, or *simulation*, is the subject of this book.

Simulation, in the sense we use the word here, is a modelling process where a dynamic reality, either actual or projected, is imitated in terms of computer actions. Today, many activities can also be broadly regarded as involving a modelling procedure. The painter, sculptor, dramatist, physician, physicist, statistician, engineer and economist all depict their own concerns by using some kind of *model* in terms of which real phenomena can be interpreted and hypotheses can be formulated. Making use of models, a practitioner can interpret particular occurrences in terms of general forms, and thereby giving a meaning to new phenomena. As a discipline matures, its established models (paradigms) are continually being compared, improved or discarded, according to their usefulness.

In fact, human consciousness can be thought of as being largely based on a modelling process. Using thought-models, we develop a framework of expectations which enable us to reconstruct reality in a coherent way, upon which basis we can make decisions, experience emotions, or pursue actions. The model is the matrix for the piecing together of otherwise disconnected experiences. Interpretation thus presupposes a model of some kind. In cognitive science, model-building is taken to be an essential part of human consciousness (Johnson-Laird and Wason, 1977), where the effect of an action is prejudged in the mind's eye. The basis of communication relies on the acceptance of a shared model between the communicants (von Foerster, 1980), by which they can formulate and interpret messages. The model mediates between the syntactical pattern of the signal and the semantic value of the message.

When we consider exactly what a model is, that is, its significance in most general terms, we are faced with a duality. From science we are familiar with models of physical reality which claim to reflect and predict real events and phenomena. When we have a good model, we can say with some degree of confidence that we know 'how' a part of Nature behaves. But, in other situations, we may use the word 'model' in the sense of an ideal form or paragon. We talk of a model student or a model factory as something that is a desirable goal, exemplifying our idea of the superlative.

Correspondingly, the practice of *modelling* can be effective in two ways. We may be able to *describe* or reflect a particular aspect of the real world, and as such modelling plays a fundamentally explicative role in science. Thus we may say that Newton's law of gravitation is a good model, according to how well it accounts for gravitational phenomena in the real world.

Alternatively, a different kind of model arises when the modeller is concerned with demonstrating how the world *should* be. Artistic visions and political convictions come into this *normative* category of models. Usually they represent an ideal to be aimed at, or imitated. Their very self-evidence seems to stir their supporters to interpret the world through these ideals, or even to try to *impose* these ideals on the world. For instance, the latest creations of a fashion house may inspire people to adopt a particular 'look', and we may judge the model's goodness by the extent to which it is accepted by the public.

The word 'model' thus has two distinct aspects. A model may reflect reality, impinging on our ideas, or may serve as an ideal towards which we want to modify reality (Fig. 1.1). Can the aspects be reconciled?

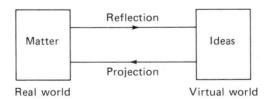

Fig. 1.1 — Idealist and materialist views of models: a descriptive model reflects reality, whereas a normative model is an ideal projection

In any situation where modelling is called for, one aspect usually dominates the other, but never exclusively. The adoption by the scientific community of the current scientific paradigm (Kuhn, 1970) certainly plays some part in the formulation of scientific hypotheses, and thereby the ideal preformulation influences the specific mode of any model in which reality is reflected. On the other hand, the seemingly 'free' creations of Art, such as painting, fashion, etc., despite their often somewhat bizarre creations, must operate within a confining model of physical feasibility, not to mention issues of commercial or social acceptance, even though a particularly bold creation may challenge such barriers. Systems theorists have stressed the separation of natural from man-made systems (Varela and Gognen, 1978).

The predominant aspect in any particular application is determined by the extent to which models can feasibly be *implemented*. In scientific activity, the experimental set-up is designed to preclude the observer from interfering with the phenomena under study. He must passively observe phenomena and sharpen his view of them by contriving different forms of experiments. The laws of Nature are not subject to his control, but await revelation. But when questions of *design* are paramount then our ideal model of a system's behaviour, which may be expressed as desired performance criteria, is to be translated into reality.

Having described our epistemological starting point, we may now introduce some terminology. There are three basic components in the modelling process: the model, the object system it refers to, and the modeller. The *model* is a representation of the object system, contrived by the modeller for some purpose. The *object system* is what the model represents, its *referent*. It is often called the *real world,* even when it might be a hypothetical variant of the real system under study. The third essential component is the *modeller*, who constructs the model as a representation of the object system, employing a *transformation* of the object system, either into another medium or into another dimension. The model can be used as a basis for discussion within a design team, a demonstration of hoped-for capabilities, or a subject of experimentation. To complete the design cycle, the experimental results are translated back into reality.

The advent of an increasingly complex and interconnected man-made environment has lead to problems of the design and integration of artificial systems. Examples of such *designed systems* abound in modern life, including all types of services, travel facilities, hospital admissions, libraries, industrial plant, banking, etc. Such systems are today made practicable largely through the existence of high-speed computers and communications facilities. In this situation the modeller is no longer eagerly seeking revelation, like the experimental scientist, but is active in the design of a system to produce a service according to a prescribed performance criterion. Many engineering problems are also of this type.

The essentially tripartite nature of modelling has been mentioned in Minsky (1968), who states that an observer B would consider an object A★ as a *good model* of A if the answers A★ gives about A are the same as those of A. For our purposes, it is not sufficient just to consider the outputs of a model: in simulation, we are interested in the evolution of systems in time, so we try to make models valid by proposing internal mechanisms. There is a further problem with Minsky's view: if A does not exist, how can we validate the model? Ultimately, all we can do is to appeal to some kind of consistency in the universe, and tentatively extend our experience to cover the gaps in our knowledge. When we consider the model as a design for a new system, we have the situation where Minsky's A does not exist. Then the modeller's power and interest in implementation of the results of the investigation cast him in the role of a manager, operating within certain *objectives*.

Discrete-event simulation operates on the intersection of the two aspects of modelling. The simulationist is therefore interested both in a valid representation of the object system and in the ability to tinker around with the model to test it under various assumptions, before committing the design to implementation. Simulation's field of application is designed systems. Consequently it has grown in importance with the increasing prevalence of designed systems in society. As a part of Operational Research (OR) it stakes a claim in the 'depopulated no-man's land between mathematical theory and practice' (Rivett, 1980).

Like the scientist, the OR practitioner using simulation is certainly concerned with the validity of his model. But in addition, the objective system has a specific function with respect to a workload by which it can be assessed. So the ideal functioning criteria of the system can be identified as a factor external to the system itself. For instance, an airport facility will be planned in accordance with projections of the future demand on its services. In trivial cases, such a view may lead to a superficial optimisation of the system, but in practice, purposes and workloads are multi-level concepts which do not allow reduction to a simple numeric form. When simulation is used for more complicated systems, the internal structural properties of the system assume greater importance.

By making detailed empirical observations on the reality under study, a more precise model can be built. The reality, suitably amended with respect to any hypothesis the modeller might wish to impose, can then be put through its paces and subjected to evaluation, based on desired performance criteria. Effective simulation thus demands both an accurate portrayal of real-world phenomena, and the need and ability to put amendments into practice.

It is in the area of model construction that the modeller's skill and art are especially required. The appropriateness of the transformations depends strongly on the overall goal of the project, and problems of accuracy and relevancy of representation cannot be resolved without reference to the *purpose*, the goal of the modelling enterprise. Moreover, purposes have an element of one-sidedness about them and depend on perspective. I might look for a barber's where I don't have to wait, but for the barber-manager, continual queueing represents 'maximum use of resources'. A more perceptive barber-manager may see long queues of waiting customers and consider the baulking effect on other potential customers and so employ more barbers. Thus the manager's role is not that of simple optimiser: the reality must be grasped as a totality, a system.

Of course, no model is perfect, in the sense that one should never expect a model to be able to encompass the totality of the real world. We expect models to be considerably simpler than the phenomena being portrayed. In fact, the very act of simulating can have a resolving effect, by forcing the manager–modeller to get better acquainted with the logic inherent in any complex system, guided perhaps by the rigour of a computer language.

Since every model involves some kind of transformation from the real world, we can say that a simplification, an idealisation, and, cynically, a falsification are involved (Turing, 1952). While the model is a safe haven for experimentation, being more convenient, more controllable and more malleable than its referent, the modeller can never be totally sure, because of the transformation involved, that his findings will transfer back into terms of real implementation. We can only hope that this uncertainty is a small price to pay for the ability to predict beyond the current situation.

## 1.2 DEVELOPMENT OF SIMULATION PRACTICE

The main concern of this book, discrete-event simulation, has developed mainly as a technique within Operational Research (OR), which can be described as the art of modelling managerial problems. Of all the techniques which shelter under the OR umbrella, simulation is the most applicable. Shannon (1975) cites a report which showed that, of the top 1000 US firms surveyed, simulation was the most frequently employed technique for corporate planning. To look for the origins of our subject, we should start from the emergence of OR.

After the rapid advances in science and its application during the Second World War, both in warfare and in industrial production, a more contemplative attitude pervaded the scientific community. The encouragement by enlightened management of research and development attempted to apply the successful managerial techniques developed in wartime to the rebuilding of peacetime economies, especially in capital-intensive industries, such as steel works (Checkland, 1983). Most of the early researchers in simulation developed their skills in steel companies.

The new approach demanded a means of dealing with the concept of a dynamic *system*, whether in the form of an economy, a workface, or a factory. No off-the-shelf mathematical structure could be found which would be suitable for such a complex of interacting components, each with their own particular characteristics. One way of making a detailed representation of a system is to build all the details into a computer program, which could be driven by a mechanism to advance time, an approach which led to discrete-event simulation. Owing to its intuitive appeal and lack of abstraction, simulation gradually gained acceptance as a way of representing designed systems. As a part of this trend we can see the increasing importance of the idea of a model as a fit subject for speculation.

While the idea of manipulating computer models is widespread and familiar today, the central role of models in formulating new ideas was not always so explicit. Traditionally, scientific insight into natural phenomena had been gained by putting forward a set of equations as a hypothesis, but the intensive scientific investigation which took place during the Second World War increased the acceptability of modelling as a development tool.

Naturally, the post-war industrial revival tried to take advantage of the new advances in the application of science which had proved so successful.

Rosenblueth and Wiener (1945) attempted to collect these approaches in a philosophical assessment of wartime developments. They set out to categorise the kinds of models used by science by splitting models into two categories: *material models* which are based on a physical transformation of objects and *formal models*, or symbolic assertions of idealised situations. Nowadays, the separation seems obvious enough, corresponding to a division of the field into models as reproductions of reality in experimental form, on the one hand, and mathematical modelling, on the other. In fact, the separation has led to two overlapping schools of thought on the practice of modelling, based on the level of abstraction adopted by the modeller. Formal models tend to be more abstract in nature, whereas material models represent problems in more immediate terms.

Material models are generally used for investigating the properties of materials or of a particular structure, and are scaled models of reality. The physical resemblance is strong, but the scale may be different. The shape of a bridge structure can be tested by scaling down; the training of aircraft pilots can be undertaken by building an artificial environment which behaves just as a real system would, whereas chemists' ball-and-stick models of molecular structures are scaled-up material models.

Formal, or symbolic, models operate on a more abstract and idealised plane. They usually involve mathematical modelling, which represents reality as a set of equations. Often a complete solution for the particular case at issue is unobtainable, except for the simplest cases. Many mathematical models are capable of translation into a form which can be solved by numerical methods which are more widely applicable.

Wiener's later instigation of the science of cybernetics (Wiener, 1948), with its emphasis on control and communication through the concept of *information* and feedback within a system, directly focused on information as a representation of a state of a model, or system. During this periods there was a general movement away from mechanistic concepts towards more system-like, holistic concepts, making system-models more acceptable. The totality of scientific endeavour could then be seen as a process of investigating models so as to develop an understanding of a system's output under differing internal conditions, as opposed to just observing input–output correlations. Contemporaneously, there was a parallel development in psychology away from naive behaviourism based on the correlation between stimulus and response, towards a more overall perspective, taking into account internal states.

Simulation, meaning a realisation of a physical law or theory in a particular circumstance, assumed more importance with the development and increasing availability of the electronic computer. It had been known for a long time that, when expressed in terms of differential equations, many different physical systems would show analogous properties. For example, a diffusion problem may have a thermal analogue. The *analog computer* developed by exploiting these relationships, enabling a general differential equation system to be explored in terms of a system of inter-connected electronic components. There is a direct correspondence between the elements of the equations and the components of the electronic circuit. Thus the idea of a computer 'configuration' representing a particular objective system of an electrical or mechanical nature had already gained wide currency in the sicentific/engineering community.

The concepts of simulation and model and their use in OR were brought together by Churchman *et al.* (1957). They consider three categories of models:

(a)   *iconic* (scaled) models;
(b)   *symbolic* models requiring mathematical or logical operations;
(c)   *analogue* physical transformations.

Later, Crosson and Sayre (1963), in a philosophical study of modelling and computer intelligence, put forward a threefold categorisation of models, showing for the first time the impact of *digital-computer simulation* on modelling philosphy:

(a)   *replications*,   or   physically   similar   objects   suitable   for experimentation;
(b)   *formalisations*, or symbolic representations based on logic or mathematics;
(c)   *simulations* in which 'symbols were not manipulated entirely by a well-formed discipline', by which was meant that an amount of randomness was allowed in the model.

By way of justification of the simulation category of models, the authors refer to the representation of the movements of a lift in the presence of randomly arriving demands for its service. The guidance of the model is by certain *policies* (such as uni-directional movement, unless service demands are absent in the present direction of movement), requiring the representation of flexible *algorithms* to represent the behaviour of the lift in response to randomly arriving demands for service. This kind of 'policy specification' could be simply specified in program form as part of a discrete-event simulation.

Many authors describe simulation as a form of Monte Carlo method. In fact, both involve programming on digital computers and share the use of random-number generators, but that is all. We prefer to regard them as quite separate pursuits (Rivett, 1980): a Monte Carlo method is best regarded as a numerical method dealing with the generation of random numbers to estimate an integral, whereas a simulation is a particular kind of model of a system involving time advance, in which random numbers may be used to stimulate uncertainty or non-determinism in the model.

Crosson and Sayre (1963) recognised simulations as a subcategory of models, of either a continuous or a discrete nature, according to which 'type of computer was employed — analog or digital'. The digital computer was pulsed, and advanced through its operations step by step. By this time, digital-computer simulations, able to represent policies in an algorithm as well as being able to deal with random components, had become, alongside analogue simulations, an accepted approach to modelling, as demonstrated in Tocher's *The Art of Simulation* (1963), the publication of which marked the emergence of discrete-event simulation as a discipline in its own right.

Geographers Haggett and Chorley (1967) have taken the classification of types of models further. Since geography covers such a wide expanse of study, bridging between the Arts and Sciences, many different types of models are required to cope with such widely varied system-like phenomena ranging from weather systems to transportation planning. Their definition of a (descriptive) model is worth repeating here:

> A *model* is a simplified structuring of reality which presents supposedly significant features or relationships in a generalised form. Models are highly subjective approximations in that they do not include all associated observations or measurements, but as such they are valuable in obscuring incidental detail and in allowing fundamental aspects of reality to appear.

Models can be categorised in several ways. We have already seen the dual descriptive and normative aspects which models may possess. They may also be static or dynamic; they can be described in terms of the material from which they are made, or the kind of abstract nature they possess: theoretical or symbolic, mental or conceptual. Physical models may be iconic (scaled) or analogue; mathematical models may be deterministic or stochastic. Models may also be classificatory schemes (as is this one: a 'model of models'); a model with even broader significance may sum up a paradigm, which is more a pattern of activity to which researchers conform. We may simplify this rich classificatory model and consider just five main qualities: symbolic, iconic, classificatory, analogue and paradigmatic, to which other descriptors: static, dynamic, stochastic, etc., may be applied.

The relation between a model and a simulation depends primarily on the extent of abstraction which the analyst wishes to apply. For mathematical modelling, an overall principle is put forward as a model, typically in terms of differential equations. The model can then be supported by realisations of it, which are termed 'simulations', looking at specific occurrences or applications of the principle, perhaps involving sampling from distributions. If the model leads to intractable mathematics, which frequently occurs, the simulation can still proceed. Simulation can be used when other analytic methods break down, or are impractical, and has gained an unfortunate reputation of being a 'last resort' method, to be applied when all else fails.

Maynard Smith (1974) uses the purpose of the model to distinguish mathematical descriptions of ecological systems made for practical purposes, which he calls 'simulations', from those whose purpose is theoretical. The value of the simulation increases with the amount of particular detail incorporated. The mathematical descriptions, called 'models', should include as little detail as possible, but preserve the broad outline of the problem. A model should retain only the essential general characteristics of the situation. Simulations can then be seen as applications of models closely adapted to particular circumstances. The flexibility and capacity of digital computers make them very suitable for this kind of expansion.

Simulation serves as a direct means of observing the hypothetical system's behaviour under a variety of conditions, with a chosen initial state. A simulation run can be repeated as often as required to establish confidence in its predictions. There are generally fewer abstractions and assumptions than are required by a mathematical model. Stochastic models can be used for component parts of the simulation, enabling a more comprehensive view of a complex system than that offered by a mathematical model. There is no 'solution' in the mathematical sense; what is gained is an understanding of the relationship between the components of the system, and some feel for selected aspects of its average behaviour.

The idea of building a model of reality and then performing experiments on the model is probably most explicit in Operational Research, although it is now recognised to be prevalent in all scientific endeavour, so much so that 'mathematical modelling' has attained its own identity as a branch of mathematics. Simulation is at an advantage by not having to rely on mathematical assumptions such as linearity, and smoothness of change, and is therefore more important in soft sciences where mathematics is not so extensively applicable (Deo, 1979).

Aris (1978) comments that the development of a model leads to further conceptual progress, whereas a developing simulation can lead to clearer understanding. Ignall and Kolesar (1979) raise the interesting point whether concentration on simulation for the analysis of queueing systems might not detract from the *insight* which comes from a probabilistic approach. Quoting Hamming (1962), 'the purpose of computing is insight, not numbers', they suggest that paying heed to this maxim will lead to simulation becoming a more important tool in systems analysis. The early approaches of OR concentrated too much in putting forward a formal model, and not enough on experimentation, whereas later approaches contained an over-reaction to this bias. They conclude that some kind of symbiosis of the two approaches would be preferable.

Computer modelling covers a wider field than models considered under the term 'mathematical models'; they may encompass masses of facts, as in a database, or at higher level, an expert system which includes factual relationships. We make a distinction between these models and simulations by defining a simulation as a computer model with a capability for *time advance*. While some databases and expert systems may be concerned with changes with respect to time, this is not essential, and can involve some difficulties. Simulations are thus *dynamic* models implemented on a computer.

Originally the word 'simulate' meant to feign or sham and had a rather unpleasant implication of counterfeit or duplicitous behaviour. In its technical meaning which has evolved within the field of OR, it has come to mean a specific kind of modelling in a computer-implementable form. Taking a broad enough view, all programming can be seen as modelling of some sort. But to be a simulation, there must be some modelling of how the system will develop in time.

The practice of simulation is thus an experimentational technique; a completely rigorously defined, predictive calculus is abandoned for a trial-and-error methodology in which policies and capabilities can be assessed. It is in this form that simulation impinges on the manager, as an assistance in evaluating options which may present themselves. The manager can test out his ideas in a 'what-if' manner, build a model of the reality incorporating an hypothesis and subject it to experiment. To estimate the state of affairs pertaining in the future naturally requires a time-development algorithm to motivate the model; the generation of randomness is of secondary importance.

## 1.3 SIMULATION AS COMPUTATION

The origins of computer simulation are closely intertwined with the early development of the computer itself. A computer program can be seen as an automation, or simulation, of the processes which a human computer would apply when performing a calculation. During the Second World War one of the main spurs to the development of automatic computation was the problems involved with coded communication, especially the decryption of intercepted enemy communications. A coded message must 'simulate', or give the impression of, a chaotic sequence of characters, yet enable a mechanism to decode and 'make sense' of it (Hodges, 1985).

How could one tell if a sequence were *really* random, or just apparently so? Clearly if the sequence passed all the tests for randomness, then it must be acknowledged as random. In Chapter 5 we will describe in more detail methods for simulating randomness by deterministic algorithms. By applying an analogous argument to the concept of 'intelligence', Turing (1950) was able to arrive at his *imitation game*, a practical definition of intelligent behaviour in machines: if a machine passes all the relevant tests for intelligence, then it must be accounted so. The possibility of machine intelligence, especially in regard to chess-playing, exercised the minds of computer designers during the 1950s. If a machine can simulate intelligent behaviour, then surely it could be made to imitate the actions of designed systems.

At the same time, OR workers were attempting to restructure and modernise post-war industries by applying their wartime experiences. OR, together with the associated disciplines of cybernetics, systems analysis and management science all faced the common problem of how to represent a *system*, whether it be a production plant, a market, or an economy. The dynamic representation of industrial processes, especially where production scheduling and planning processes for large systems of capital-intensive plant were called for, was an area where computers could assist, since a computer simulation of a system would enable experimentation with various factors to be performed without disturbing the real system. Simulation thus forged a link between contemporary OR and computing developments.

The steel industry was particularly in need of this kind of help and many of the first applications of simulation occurred in this field. In the late 1950s and early 1960s, Tocher laid much of his groundwork for computer simulation for application to steel-production plants, in the course of which he investigated random-number generators, data structures for entity description, search control structures, etc., but above all he developed a linguistic means of representing the flow of entities through the plant, in his 'machine-based' approach to simulation, aided by 'wheel-chart' diagrams. In addition, he was also concerned with several peripheral areas such as interactive computing, run-time facilities, as well as embarking on building his own computer from war-surplus materials (Rivett, 1983). Elsewhere, the first versions of high-level simulation languages GPSS and SIMSCRIPT were being developed.

In the early stages discrete-event simulation, as described in the first book on the topic (Tocher, 1963), the Monte Carlo associations of simulation were still very evident, with language developments shyly confined to final chapters. But one year later, in a very perceptive paper, Tocher (1964) gives a more complete description of the simulation language, GSP. Basically, the production-plant system is seen as a collection of interconnected queues, with cooperation between machine cycles. As a browse through the paper will reveal, the language has a decidedly 'machine-code' look to it, as compared to GPSS and SIMSCRIPT, with single-letter identifiers and jumps, necessitated by run-time efficiency which was at a premium.

Enough simulation languages had been proposed by the early 1960s to require a comparison of them, with a future projection of their developments (Krasnow and Merikallio, 1964). Discrete-event simulation issues were one aspect of the general questions being discussed at that time on the future of computer languages. How should dynamic methods of memory management be employed? What should be the next stage in the replacement of Algol 60 as an all-purpose algorithmic language? For simulation, the most significant development was the emergence of Simula to create an Algol-like programming language for simulations, a development which originated the object-oriented approach to programming. A few years later, a conference devoted to the problems of simulation programming languages was held (Buxton, 1968), by which time the inventiveness seems to have reached its peak.

For all types of computer languages, whatever the application, there was a general move to democratise computer programming away from a programming elite, mystically fluent in machine code, which was based partly on the realisation that programming must be de-skilled and more amenable to non-expert use, and partly as a way of ensuring that programs should be correct. There were also other considerations. Because computer designs were being continually updated, use of a language defined too close to the hardware lelvel would make it difficult to transfer programs to new machines.

The movement culminated in the development of higher-level languages, nearer to natural language. Not everyone sympathised with the new development. Tocher (1969) thought the trend toward natural syntax would 'hinder reconciliation to the necessary formality', the only people to gain being the partially interested. The movement met with general success, however. But it soon became clear that the newly proposed languages were not sufficiently general purpose to allow simulations to be conveniently written.

The arrival on the scene of Algol 68 alleviated some of the problems, supplying dynamic data management and data structures, but without run-time binding of identifiers or abstract data types. It remained awkward to write simulations in Algol 68 (Shearn, 1975), although attempts to do so were seen as particularly revealing of the language's weaknesses (Levinson, 1973, 1974).

The global assumption that segments of program should be strictly blocked into a neat hierarchy, put forward to discourage the messy logic which arises from flagrant use of the **goto**, conforms with the mathematical substratum of computing, along with provision for nested arrangements of bracketing, and nested calls of functions. With the advent of Structured Programming, hierarchical structure emerged as the tacitly accepted norm, in the sense that any deviation from this pattern was seen as error-prone.

While mainstream computer programming continued to be centred around the automation of mathematical computations, a simulation program does not necessarily fit well into this structural context. The flow of control of a simulation program basically represents the logic of state-changes in the system being pursued with the elapse of time. These flows can be repeated to represent replications of event sequences, the performance data from which can be aggregated to provide output statistics on system behaviour. For scientific calculations, the control structure by which the conclusion is arrived at can be altered as long as the result remains correct; for a simulation, the control logic should in some sense 'mirror' reality, and 'correctness' can be assessed only insofar as the reflection of reality is valid.

The maxims of a hierarchical program structure are not suitable for simulation by coroutines, which are program segments which can be entered and exited at places other than the start and end of the segment. Coroutines, when entered, can make use of information retained from previous invocations in the form of values of local variables. Even when coroutines are not used, it may be necessary to invoke procedures defined at run-time, or to simulate entity flows which interrupt one another, which would again transgress the control-flow hierarchy. In simulation, the strict hierarchical format of Structured Programming may need to be breached.

The lack of conformity with structural norms may have led to the slackening of interest in simulation during the 1970s. During this period the simulationist was confronted by a plethora of languages and simulation methodologies, but few really new ideas. Stagnation was partly due to the awkward position occupied by the discipline: as a branch of OR, it is excessively pragmatic and computer-oriented, lacking any reassuring mathematical back-up, whereas as part of computing, it does not fit well into the accepted form of a hierarchicaly decomposable program of computing primitives. Mathematically, it has often been regarded as a discipline of 'last resort', only to be considered when the orthodox approaches of queueing theory have been proved intractable.

The detrimental effect of commercial factors can be seen in the lack of wide acceptance of Simula (Virjo, 1972). As a genuinely general-purpose language, Simula can embrace both calculational programming and simulation, including abstract data types and general package-writing facilities in the form of the object concept. But only after 20 years is the language beginning to receive the proper attention that it is due. However, as we shall demonstrate, Simula is not the last word in simulation, embodying a particular strategy in the programming of simulations, which leaves it open to certain errors of omission.

The design of a language for simulation should have the ability to embrace all conceivable situations — it is too easy to provide *ad hoc* solutions in the form of a new construct for every new situation. Thus the designer is led to an investigation into the complexity of structures with which the language is to be expected to cope. Simulation languages and complexity are surveyed in Chapters 6 and 7, whereas in Chapter 8 we consider the approach to designing a new language by defining a semantic basis for language constructs.

In this book, we emphasise the role of simulation language in making possible the description of dynamic systems. Language constructs should encourage the simulationist towards writing a valid representation, drawing attention to points which may have been overlooked, and allowing separate logical paths to be defined independently. The overall structure of a simulation program is more reflective than normative. Inevitably, the limits of a computer language will put bounds on our simulations; just as natural language constrains thought, so we should be looking at ways of extending the power of language for simulation.

# 2

# Time advance

Having established that the essence of simulation is the modelling of a *dynamic* system, i.e. one which evolves in time, our first consideration must be how to represent the forward movement of time, as a backdrop to the changes of state in the representation of the object system. For all types of simulation, the mechanism of time advance is the prime mover for all other effects. Selecting to study systems which evolve in time does not mean that the applications of simulation are free of the problems which beset other types of computer models, such as databases, however. We will see later, especially in the study of entity allocation in Chapter 8, that simulation shares many of the problems of databases involved with marshalling entities with attribute values, though perhaps in not so severe a manner.

To represent time flow in a digital computer, we may use a numeric variable to store its successive values. Whether we choose an integer or a real value for time depends on the type of time advance we envisage for the simulation — basically it depends on whether we want to constrain time to a discrete set of points in a temporal continuum, or whether we want to allow every point in the continuum the possibility of being an occasion for a state-change. This chapter considers different assumptions about time and their effect on the simulation of dynamic systems.

Is time really continuous, or is it made up from discrete quanta, like energy? People's views on the nature of time seem to depend on the use they intend to make of it. Goldstine (1972) relates the discrete–continuous controversy to the history of the development of the computer. While lengthy computations had been required for a long time for the calculation of tables of astronomical data and ephemerides, as late as the 1940s physicists had been content to see their computational investigations carried out in an analogue, and therefore necessarily continuous, fashion. It is reported that even Wiener, the 'Father of Cybernetics' held this view, against the convinced discretist von Neumann.

With the emphasis of this book being on the *structural* aspects of simulation, this means that we concentrate on the identification of dynamic form, irrespective of the particular context in which the form is expressed. Our investigations naturally lead to a means of expressing form, i.e. a language with constructs for form abstraction. The temporal representation of a dynamic system may be couched in either continuous or discrete terms, and it is interesting to note that such duality of description is not confined to forms of expression involving computation. The question of the corpuscular or wave nature of light has exercised the minds of physicists for many years, and, to give just one example from art: the first decades of this century saw the contemporary battlelines drawn between Marinetti's flowing dynamism directing the Futurist manifesto of art and the discrete fragmented planes of Cubism.

Of course, whichever basis we assume, there is no absolute barrier between continuous and discrete phenomena: continua may give birth to discrete effects. For example, weather-systems may spawn cyclones; rain-drops may coalesce to form rivers; flowing streams may give rise to vortices; and so on. These phenomena are the focus of concern for mathematical catastrophe theory (Thom, 1975), which attempts to characterise the forms of discontinuities arising from continuous flows.

## 2.1   DISCRETE-TIME SYSTEM

In a discrete-time system (e.g. Cadzow, 1973), the model advances to each successive stage in a series of jumps, the time being incremented by fixed amounts. After every 'clock tick' the value of time is increased by a constant, leading to a *synchronous* time advance. One problem in implementing a simulation of a synchronous time advance is that all time-dependent happenings occur at instants when other happenings are also taking place, so we need to implement a fictitious state-space in which to store the intermediate results of the state-changes, before committing the changes to actuality. The states of the model are then necessarily discrete, persisting between instants, although if they are close enough together they may approximate a continuum.

The fixed-increment discrete-time system simulates time advance by a method similar to the way motion is represented in a cinema film. A motion picture consists of a reel of film containing a series of still scenes which are projected on a screen at a rate of 24 frames per second. The rapidity of the projection and the correlation between successive frames deceives the eye into interpreting the whole effect as continous motion. Along with the general illusion of motion, there are also some effects which can be somewhat disturbing — most cinema-goers attending a Western film are familiar with seeing wagon wheels appearing to go into reverse. Any kind of moving image of an object possessing self-similarity, e.g. the regularity of pattern in the spokes of a wheel, can be susceptible to stroboscopic effects, and similar problems can beset fixed-increment time advance.

Many designed systems operate with fixed-time increments. Bank interest is compounded regularly, whether monthly, quarterly, half-yearly or annually. The choice of the interval will have an effect on the total interest raised. Discrete-time systems, in which time is advanced by a fixed increment, are commonly used to describe dynamic systems in electronics, control theory, and communications. Åström and Wittenmark (1984) give a description of computer-controlled systems based upon this form of time advance. Although functions of time are allowed to be continuously changing, in order to accommodate cumputer control, the time-varying functions are only partially known through a *sampling* procedure applied to them. The sampling rate effectively limits the responsiveness of the system, by assuming that the range of frequencies in the signal is bounded, through the Shannon–Weaver Bandwidth theorem.

The simplest example of the basic discrete-time model is the *production system* which defines a grammar. It is defined by a set of productions A→B, C→D, etc., where the left-hand side of the arrow → is replaced by (or *rewritten* as) the right-hand side. Each application of a rule can be thought of as an advance in time, but is often just a stage in a production sequence. Syntactical definitions of languages can be described by such a grammar. A slightly more complex example of a production system is a Turing machine, which is a computing machine reduced to its simplest abstract form, and is the basis for the von Neumann conception of the digital computer. Symbols can be read from and written to an infinite tape, and the machine possesses a set of internal states by which the particular production can be conditioned. Zeigler (1976) gives an abstract formalisation of a discrete-time system as a realisation of a sequential machine.

There is also a host of synchronous models where the stages alternate. Many games are built up from steps like this, where each ply, or change of state caused by one competitor, must be made in proper sequence. Some games, such as chess, and the Chinese board game *wei ch'i*, in which each player seeks to surround his opponent's pieces, originate in simulations of battlefield situations, as we often can gather from the names of the pieces.

Board games are two-dimensional productions, with effects from neighbouring elements. One very simple game, which simulates the growth of a colony of 'cells', is called the Game of Life. Invented by Conway, it was publicised in Martin Gardner's *Scientific American* column (Gardner 1970, 1971). The player is presented with a two-dimensional grid and invited to define an initial configuration of cells. The configuration is 'grown' by being subject to a set of rules governing the birth of new cells, and the death of isolated or overcrowded cells, with respect to the neighbours of each cell. Each cell can be live or dead (1 or 0) and each cell position has a neighbourhood of eight surrounding cells (Fig. 2.1(a)). Notice that the neighbours of a given cell are of two types, sharing either a common side, or a common corner. The state-transition rules are:

(1) 0→1 for 3 cells in neighbourhood;
(2) 1→1 for 2 or 3 cells in neighbourhood;
(3) 1→0 for <2 or >3 cells in neighbourhood.

The game is thus a two-dimensional production system, where the productions are dependent on the number of neighbours, and is therefore a context-sensitive, cellular automaton. Other similar schemes are discussed in Preston and Duff (1984). They all share the property that, although completely deterministic, they can exhibit a wide variety of behaviour: a pattern's future development is predictable only by letting it live its 'life'. No

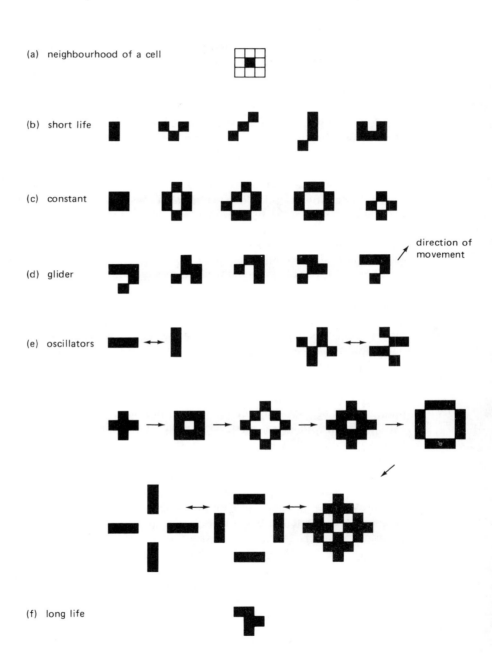

Fig. 2.1 — Conway's game of life: some patterns.

real time-advance *mechanism* as such is really required except the simple updating of a variable to indicate the generation number, since, being a synchronous form of time advance, something happens on every time increment.

The interesting result is that although the program, is completely deterministic — apart from the choice of the initial configuration of live cells — it is so surprising in the way each pattern of cells develops. It is difficult to predict the qualitative results. Conway was seeking a set of rules with bounded growth and stability, yet we can observe five types of evolution: a pattern may die out, remain constant, undergo translation, oscillate or continue growing for a large number of generations. Some example patterns are shown in Fig. 2.1(b),(c),(d),(e),(f).

The simulation is not intended to be a genuine simulation of a real colony of cells, but an investigation of the abstract types of phenomena which may occur in the simplest of circumstances. The evolutions are quite sensitive to a change in the rules, and to the definition of the neighbourhood of a given cell. Translating the program to a hexagonal grid means that all six neighbours are of a single type, and the production rules must be amended accordingly. Whichever way this is achieved, new sets of patterns with different properties are obtained.

Another context-sensitive generative grammar, this time one-dimensional, but with *parallel* rather than *sequential* rewriting, attempts to match more closely the natural situation of biological growth. The theory of Lindenmayer systems (L-systems) was first described in Lindenmayer (1968), the original aim of which was to provide mathematical models for the growth of simple filamentous organisms. Abstracted from its biological motivation it has been vigorously pursued as a branch of formal language theory. The basic postulate is the idea of *parallel* rewriting. If a grammar contains the production rule S→SS where sequential rewriting is assumed then any string of the form $S^n$, for $n \geq 1$, is obtainable. But in an L-system we can obtain strings only of the form $S^m$, with $m = 2^n$. This reflects the basic biological system behind L-systems in which the development takes place simultaneously everywhere, and sequential rewriting is not applicable.

The Game of Life and Lindenmayer systems show that a simple deterministic rule-bound dynamic can produce a surprisingly complex variety of outcomes. There is another deterministic system, exhibiting complex behaviour, consisting of a recurrence relation which is even simpler in its definition. The basic relation, the well-known *logistic* equation of population dynamics, may be expressed (Aris, 1978) as

$$x_{n+1} = \lambda x_n (1 - x_n),$$

where $x_0$ is chosen from the interval [0,1]. The transformation maps the interval into itself, for

$$0 \leq \lambda \leq 4,$$

and the constant $\lambda$ may be used to 'tune' the behaviour. As $\lambda$ is increased from zero, the origin is the only fixed point, until $\lambda$ exceeds one, when another fixed point emerges at

$$x=1-\lambda^{-1},$$

and the origin becomes unstable. When $\lambda$ reaches 3, two stable points of period two (i.e. for which $x_{n+2}=x_n$) emerge. At

$$\lambda=1+\sqrt{6}=3.45,$$

four solutions of period 4 emerge. The process of binary fission recurs with increasing frequency until a limit point occurs at $\lambda_c=3.57$. Beyond this point there are a countably infinite number of unstable periodic orbits and also an infinite number of solutions which are in no sense periodic, that is they are neither periodic nor asymptotic to a periodic solution. After $\lambda=3.83$ there are cycles of all periods as well as strictly non-periodic solutions — a situation aptly described as *chaotic*. Not only does the logistic equation display such a variety of properties, but it typifies a generic situation for all functions with a 'hump' and a parameter by which they can be tuned.

The *orbits,* i.e. sequences of generated numbers, of these simple deterministic mathematical models can explain several of the experimentally observed routes to chaos. Certain features are universal, in the sense that they are largely independent of a system's detailed structure. Chaos is formally defined (Wolf, 1983) by the property that nearby orbits show a rapid (exponentially fast) divergence in their time evolution. In numerical simulation this sensitivity to initial conditions is important, since recurrence relations affected by chaotic behaviour may diverge rapidly. In addition the existence of chaos raises the question of whether the matching of the results of a computation with experience can ever really be trusted. These, and similar, considerations lead to further unanswered questions relevant to the validation of a model.

But rather than discourage mathematical modelling, it may be that many naturally chaotic behaviours will turn out to be mathematically tractable. That even such a simple model as this should show such a variety of behaviour may well reflect the irregularities found in Nature, in phenomena such as turbulence, etc. We report in Chapter 5 an application using a chaotic recurrence relation as a source of pseudo-randomness.


## 2.2   CONTINUOUS TIME

Mathematics, since the invention of the differential and integral calculus in the seventeenth century, has worked successfully with the concept of a vanishingly small increment of time with which to model continuous-time advance. The magnitude of the increment is not fixed, but tends to zero and the time derivative is the difference between function values, divided by the time increment, taken 'in the limit'. Digital computers, being fundamentally discrete devices, in that the basic change of a value in memory is necessarily a discrete change, naturally lead to thinking in terms of modelling time advance as incrementing a variable by a small positive value, which is the basis for a *discrete-time system*.

Accepting that continuous advance is impossible to replicate in the digital computer, it can be approximated using a suitably small increment and by following the methods of numerical analysis. Successful application of both calculus and numerical analysis presumes that the time-evolving function of interest is *smooth*, i.e. it possesses derivatives of arbitrary order, and *analytic*, i.e. expressible in terms of a Taylor series which is convergent (Poston and Stewart, 1978). Gradually, it has become more widely realised that few 'naturally occurring' curves are really analytic, which can be indicated by their possessing non-integer dimension (Mandelbrot, 1977).

The mathematical solution of differential equations is only possible in a few cases. Success depends on the system being expressible in terms of a function which is amenable to integration by the standard techniques of calculus. Many practical problems cannot be solved at all, or have cumbersome solutions. Normally, if a mathematical approach shows no promise, a numerical solution is attempted. Numerical solutions, being based on a finite model of the differential, involve an approximation, and it is not always easy to tell the reliability of the solution obtained.

Numerical methods arise whenever the differential is approximated using finite quantities. Although digital computers cannot deal directly with continuously changing magnitudes, differential equations can be solved by programs which are based on those numerical methods developed for hand calculation (e.g. Dorn and McCracken, 1972). To solve a differential equation

$$y' = f(x, y)$$

means to identify a function $y$ whose derivative with respect to $x$ satisfies the equation. Differential equations are by far the commonest example of models in science. Normally the equation will satisfy a family of curves, and we need an initial condition

$$y(x_0) = y_0$$

in order to obtain a unique solution. The differential equation gives us the *slope* of the curve at any point as a function of $x$ and $y$, and we know that the solution passes through $x_0, y_0$. If we follow Euler's method and compute the tangent at that point, we may advance along the time axis ($x$-axis) for a small distance $h$ arriving at $x_1$, which has a corresponding $y$-value $y_1$. This procedure may be repeated, inching along the time axis by small increments, obtaining estimates of the solution at these points $\{x_i, y_i\}$.

Clearly, there are pitfalls in Euler's method. The choice of $h$ will affect the solution, and also the speed at which the calculation can be carried out. Approximating the curve by a sequence of line segments will soon go astray if the chosen value of $h$ is too large, or the slope is rapidly changing. The *stability* problem, as it is known, lies in the fact that the estimate of the gradient used at $x_{i+1}$ continually lags behind, and any small error in $y$ will tend to become magnified at each successive step.

In order to take more account of the gradient over the range of the increment $h$, we can include the value of the slope at $x_{m+1}$. By taking the average of the slopes at these two points, we obtain the *improved Euler method*. Alternatively we can use the slope at the average of the two points: $x_m$ and $x_{m+1}$, an approach which is termed the *modified Euler method*.

Euler methods are particular kinds of *Runge–Kutta* approaches which are characterised by being first-order methods, i.e. only the information on the previous point is used to get to the next point. This means that the step size, $h$, can be readily changed to accommodate irregularities in the function, or to synchronise with an event caused externally. When numerical methods are used to simulate systems with continous and discrete parts, this property is especially useful. Other methods are available which use information at one or more additional points, and these methods generally incur less evaluation of the function. Often, methods of both types are used in combination.

The type of time advance occurring in the integration of dynamic systems defined by differential equations is thus a *discrete-step* approximation to continuous change. Depending on the type of numerical method in use, the time advance can be of the fixed-increment variety, or the interval can be locally adapted to the gradient. Systems with discrete components evolving over continuous time can thus be modelled by a discrete-time approach. Greenspan (1973) adopts a fixed-increment time advance for physical models of atoms and galaxies, arguing that since observations on Nature are necessarily discrete, then the most effective means of modelling them will also be discrete, especially when the digital computer is the modelling medium.

Both time and space may be discretised: space may be segmented by one of three methods: *finite differences, finite elements,* or *particles.* The first two are commonly used to simulate structures in engineering, with the finite element approach being more adapted to the shape of the structure being modelled, and particulate models are used for systems undergoing temporal evolution especially when the physical system is corpuscular in nature (Hockney and Eastwood, 1981). The particulate model can be seen as a temporal development of an initial-value boundary problem and can be pursued in one of three ways.

The particle–particle (PP) method denotes the state at any time $t$ by the set of particle positions and their velocities. A time-step loop updates these values using the forces of interaction and equations of motion to obtain the state of the system at a slightly later time. PP implementations are usually very time-consuming. The system may be speeded up by approximating values of field quantities on a regular array of mesh points. Finite difference approximations on the mesh replace differential operators, such as the Laplacian. With the particle–mesh (PM) method, potentials and forces at particle positions are obtained by interpolating on the array of mesh-defined values.

The PM method is much faster, but less accurate, than the PP method, and can handle only smoothly varying forces. A third method (particle–particle—particle–mesh, or $P^3M$ method) splits the interparticle forces into short-range and more slowly varying long-range forces, the short-range forces being handled by the PP method and the long-range forces by the PM method.

The idea of computer simulation started with *analog* computers, in the form of setting up an electrical circuit analogous to the (usually mechanical) system under study. Such analogies depend on the sharing of the same differential equations between the two systems — the point of the transformation is that the electrical circuit is easier to measure and modify. The analog computer is an extension of the procedure, dealing more abstractly with elements of differential equations directly. An integrator, for instance, is represented by an operational amplifier, and many integrators can be operating in parallel. Dependent variables are continuous variables, and, once the circuit is assembled, the response is immediate. Minor amendments in system characteristics can be made by altering settings of potentiometers.

Originally the complexity and response-time requirements were such that only analog computers could provide a solution. A typical task of one type of analogue computation is the accurate portrayal of an air pilot's environment. The pitch and roll of the aircraft in response to the adjustment of controls, together with a view of the horizon or an airport runway, is modelled in real time by an aircraft *simulator*. However, the accuracy of representation is limited by the quality of the circuit components. It is difficult to precisely duplicate an environment, owing to 'drift' in the characteristics of the electronic components. For a while, the idea of combining analog and digital computers into a hybrid form was pursued, but the advances made in digital-computer technology have superseded such approaches, at least for most applications. Even pilot-training simulators are now based on digital computers.

The gradual changeover for continuous simulation from analog to digital computers was accompanied by the emergence of continuous simulation languages, such as CSMP (Brennan, 1968). With both continuous and discrete simulations being run on the same type of machine, there were moves to integrate the two approaches, and to develop unified software which could not only simulate systems undergoing continuous change and undergoing discrete change, but also to simulate systems in which both kinds of time development are significant. The first moves toward a *combined* simulation system, in which the dependent variables of a model may change discretely, continuously, or continuously between discrete events, were made in the late 1960s (Fahrland, 1970; Golden and Schoeffler, 1973).

Treating relationships between variables as if they were analytic or smooth underlies most mathematical treatments of functions, including catastrophe theory. Little consideration is given to the possibility that *natural* relationships may not be sufficiently smooth to support the usual mathematical edifice. Discrete-event formalism relies on no such assumption, which leaves open the possibility of novel interpretations of dynamic systems.

## 2.3   STOCHASTIC METHODS

### 2.3.1   Stochastic processes

Most applications of probability theory in statistics start out from a hypothetical model of observed phenomena. Bartlett (1975) remarks that model building is an essential part of the application of probability, whereas deterministic theories tend to apply to the more abstract elements of a system: models are more relevant when the holistic, systemic aspects are to be taken into consideration. Stochastic models, or models involving a probabilistic element, also offer the advantage of making a statement about the expected effect over a range of input values, whereas a deterministic argument deals with only a single input value, which is not so helpful in many applications. Although realisations of probabilistic models might require more computational labour, perhaps extending over several replications, the modeller thereby receives a level of confidence about variations in the input, and sensitivity analyses can be applied.

There are many ways in which probabilistic concepts may be used. Static models may make use of *fuzzy* concepts in which basic mathematical relationships are held to be true or false according to some level of probability. In a more dynamic sense, a *statistical* model may be assumed as a signal output from a black box which is masked by a random 'noise' signal. If the output is sampled at discrete points in time, then a discrete-time system might be appropriate.

The origins of *stochastic processes* lie in the study of Brownian motion and other random-walk problems. Basic to any discussion of stochastic processes is the concept of *state*. While classical mathematical analyses have been concerned with continuous deterministic functions, probability models often involve discrete states. The state is the means by which the changes due to the elapse of time manifest themselves, and first received attention from Markov (1856–1922) in his study of linked probabilities (Boyer, 1968).

Like time, the state can be discrete or continuous. Among one-dimensional stochastic processes we can therefore distinguish four cases (Cox and Miller, 1965):

> discrete-time, discrete-state space;
> discrete-time, continuous-state space;
> continuous-time, discrete-state space;
> continuous-time, continuous-state space.

As a simple example of a stochastic process of the first type, we may consider a system with three states, and a *transition probability matrix*. Let the states be numbered $\{1, 2, 3\}$, and the transition probability matrix be

$$
\begin{bmatrix}
p_{11} & p_{12} & p_{13} \\
p_{21} & p_{22} & p_{23} \\
p_{31} & p_{32} & p_{33}
\end{bmatrix}
$$

where $p_{ij}$ denotes the probability of transition from state $i$ to state $j$. The Markov property, which is the assumption made to enable theoretical development of the situation, demands that the probability is dependent on the current state only: a Markov process has no 'memory' of what states may have been occupied before the current state.

A realisation of a Markov process can be obtained by using a source of random numbers to perform the state-change at each time point, by selecting the next state according to the probability from the appropriate row of the transition matrix. Such 'simulations' are recommended for getting the feel of a particular process, but as simulations of realistic systems, they do not allow much representation of the effects of one element on another, for instance. Nevertheless, the Markov process, with probabilistic elements and discrete states, can be seen as a primitive precursor of the discrete-event model.

Many systems which incorporate growth phenomena — for instance, models of birth-death processes, epidemics and populations — can be described by either deterministic or stochastic models. Quite often it is found that the results for determinstic and stochastic portrayals of the same system will diverge. Conolly (1981) provides an interesting comparison. McQuarrie (1967) considers the formulation of a stochastic-approach chemical-reaction kinetics, where reactions are regarded as types of birth–death processes. The stochastic approach gives higher moments, and enables fluctuations to be considered. The study of stochastic models involving spatial propagation, which can be thought of as a continuous analogue of cellular automata, is a currently active research area.

Kleinrock (1975) gives a useful categorisation of stochastic processes. The Markov process insists that the current state of the model summarises all the information upon which the next state can be defined, i.e. there is no 'memory' of previous states. A birth–death process is a kind of Markov process with a discrete state-space for which the state of the process (representing the number of individuals in the model) at any transition can change by $-1$, $0$ or $+1$ only. A Markov process can be generalised by allowing the persistence of a state to obey an arbitrary distribution, rather than being restricted to a transition on every unit time increment, giving a semi-Markov process, which is the underlying statistical model of discrete-event systems and activity networks (Elmaghraby, 1977).

The field of stochastic procesess may be considered from another perspective, however. A time-series, consisting of an ordered set of data from observations made at different points of time, may be considered to be outputs from a stochastic process. The data values may be used, appropriately weighted, to predict future values in the series. Prediction by time-series assumes a similar standpoint as the solution of a differential equation by numerical methods, in that they assume future behaviour can be discovered as some function of past data. There are a variety of diagnostic tools which can be used either in the time domain or in the frequency domain (Chatfield, 1980).

### 2.3.2 Queueing theory

Queueing theory (Kleinrock, 1975) is an elegant mathematical development of the theory of Markov processes to the study of queues and queuing systems. It has developed since the Second World War as part of Operational Research, and is used as an analytical tool in the study of computer operating systems and networks.

A queueing system can be described in a five-part descriptor of the form

A/B/$m$/K/M

where A and B are the inter-arrival time and service-time probability distributions, $m$ is the number of servers, K is the storage capacity for waiting customers, and M is the size of the customer population. The descriptor specifies the particular queue model of interest. In most cases the final two parameters are assumed to be infinite, and are often omitted from the descriptor. There is also a standard nomenclature for the probability distributions — $M$ stands for the exponential distribution and we also have $D$ for deterministic and $G$ for general. The simplest type of queueing system is denoted $M/M/1$, which means a single-server queue with an exponential inter-arrival rate and exponential service time. A birth–death process is a stochastic population model, which can be elaborated to incorporate immigration effects by assuming a constant replenishment of individuals at a certain rate. In the case where we assume a service system with a mean arrival rate, a mean service rate and an unlimited number of servers, we obtain an $M/M/infinity$ queueing system (Conolly, 1981).

The basic elements of a queueing system consist of a flow of customers into the system, each of whom is served by one of a set of servers. When there is an excess of customers over available servers, the customers must wait in a queue for a server to become available. Customers arrive with a particular inter-arrival probability distribution, and the service duration for each customer is determined by the service-time probability distribution. From a statistical point of view, there are thus two models, arrival and service, which are linked together by the queue of surplus customers.

The *state* of the system can be effectively defined as the number of customers waiting, or queue length. Once customer and server get together in the serving activity, the customer is considered to have left the queue and the server adopts the busy (i.e. unavailable) state. When the service is over, the server becomes available once more for other members of the queue, and the customer, whose service is complete, quits the system.

The whole experience can be viewed from either the customer's or the server's viewpoint. The customer arrives, waits until a server is assigned, receives service and leaves, whereas the server's role is a simple alternating circle of being busy, then idle. To the customer, the system is resource-limited if the number of servers is not in excess, and to the server, if the customers arrive so frequently that their average arrival rate exceeds the capability of the servers to serve them, then the system is overloaded. From the management point of view, an acceptable match between level of service and arrival of customers is usually desirable.

The mathematical development describes a queueing system in the form of a set of differential-difference equations, reflecting the combination of continuity in time and discreteness of state-space. Having specified the system, queueing theory can answer some questions about its features. We may be interested in the expected waiting time, average number of customers in the system, or the length of continually busy or idle periods. Another important feature from the management point of view is the server utilisation, the average proportion of time spent in serving customers. Although the approach is useful theoretically, as a practical model for understanding real systems which involve queueing, queueing theory often requires unrealistic assumptions and when formulated tends to be intractable. Solutions exist only for a few specific cases.

Besides the simple features of queueing systems that we have discussed above, there are also other features which may characterise a real queueing system. When a customer arrives and all servers are busy, the customer must wait according the *queue discipline* until a server becomes available. The discipline for serving customers may be first-come, first-served (FCFS), last-come, first-served (LCFS), random, or according to some priority attribute attached to the customer. In the first two cases, the disciplines are implicit in that the order in which customers take up their service is solely dependent on the order of arrival; these cases may be equivalently described as first-in, first-out (FIFO) and last-in, first-out (LIFO), respectively. In computer science, a FIFO queue is generally called a 'queue', a LIFO queue is called a 'stack', and for customers waiting with explicit priorities, a 'priority queue'.

Apart from the queueing discipline, there are further features of customer and server behaviour which might be relevant. We may have different classes of customer, with their own inter-arrival and service-time distributions; perhaps they must be matched with servers of a particular type. The servers may be unavailable for certain periods; customers may defect after a certain period of time, or they may change from one queue to another (jockeying) or baulk before entering the queue or pay a cost to obtain a better position in the queue, and so on: the possibilities are endless.

However, and this is not always made clear in introductions to queueing theory, most of these phenomena are not expressible in terms of the statistical theory, which concentrates on the probability distributions of the linked models of arrival and service. Since only gross throughput is measured, the individuality of customers is not preserved. On the other hand, the simulation approach, in which the structure which links the two models is represented by a computer program, is much richer in detail and can model specific phenomena much more closely.

The main motivation for the interpretation of queueing systems by discrete-event simulation is that the purely mathematical analysis of multiple random arrivals is possible only in the simplest of cases (Hartley, 1975). Rivett (1980) comments that queueing theory is hardly an applicable theory in the sense of being able to say anything useful about real queues. However, a preference for mathematical analysis over a simulation approach persists, perhaps because the latter's use might be an admission of shortcomings in queueing theory. For instance, Kleinrock (1975) prefers to ignore both the numerical analysis and the simulation approaches since they do not lend themselves to textbook material, although he admits that they are of importance in studying real-world queues.

## 2.4　DISCRETE-EVENT SYSTEMS

### 2.4.1　Queueing-system models and structures

Discrete-event simulation was originally designed to solve complex queueing-theory problems, which arise when a designed system is interpreted as a system of interacting queues. In place of the 'customer' we have the *flowing entity*, whose route through the system defines the linkages between the queues, and for the 'server' we have the *resource*, which interacts with the flowing entity in activities. The designed system can be seen as being resource-limited, insofar as the number of resources of different types, and the periods of time when they are available, limit the flow of entities. In practice, the development of simulation is most useful when applied to systems where resource utilisation is at a premium, and is especially applicable to systems comprising large-scale, capital-intensive production plants, such as found in the steel industry — where indeed the technique found its earliest applications.

We saw in the last section that a queueing system could be regarded as a pair of statistical models interacting through a structure (the queue) in which surplus customers are stored. We also saw that since the statistical representation of the system's behaviour is limited to considerations of gross throughput, the details of customers' passages through the queue are not amenable to close description by queueing theory. In discrete-event simulation, the structure and behaviour of the queue with respect to individual entities can be made the prime focus of interest, being represented by the actions of a computer program.

The separation of these two foci has led to a divergence of viewpoints about the nature of discrete-event simulation. Statisticians tend to see the queueing system primarily as a set of linked statistical models, some authors preferring to regard simulation merely as a convenient pedagogical vehicle for teaching statistical concepts. Consequently, the algorithmic and representational aspects are often pushed into the background, or regarded as 'given' by the software.

Originally the statisticians' view of simulation as an eleaborate Monte

Carlo method predominated, but in time it has come to be realised that discrete-event systems can reflect more closely specific details of a designed system, besides its statistical properties. This ability becomes more important as the design becomes more complex. As with the Game of Life, a system does not need stochastic elements for it to possess interesting or surprising phenomena. The modern, structural, view of discrete-event simulation (e.g. Kreutzer, 1986) is more concerned with issues outside of purely statistical interest: viz. the algorithms, the methodology, the control structures, the complexities: it is on these aspects of simulation that we shall be concentrating in this book.

The technique of discrete-event simulation developed out of a synchronous, fixed-increment approach to time advance, amended to support the modelling of complex queueing systems. But it was the requirement to model random *activity times* that made a fully synchronous approach difficult to implement: either the assumed $\delta t$ would be too small, incurring slow-running programs, or the time-grain was too coarse and events which should occur might be omitted. As we shall see in Chapter 8, a similar problem recurs when attempting to implement a *wait-until* procedure (Vaucher, 1973) within a discrete-event system context.

Hogeweg (1978, 1980) has considered the effect of relaxing the synchrony of fixed-time-step methods of time advance which underlie the definition of L-systems, to systems with some degree of local synchrony and variable timing. It was found that the morphemes developed in L-systems depend more on the timing regime than the state-transition function. Local synchronisation is enough to recapture the complexity and richness of pattern and form variation developed in globally synchronous L-systems, but this is rapidly lost by asynchronisation.

Rather than assuming or approximating continuity, discrete-event simulation bases itself on a continually evolving set of discrete instantaneous events, i.e. a set of discontinuities along the time axis at which the model state is open to change. Time advance then hops from one discontinuity to the next, in ascending order. At each event, appropriate actions to model the state-change are undertaken as the event type dictates, including the addition of further future events to the set. Discrete-event simulation thus avoids the abstract symbolism of calculus and simulates directly the workings of the real world, giving the method an intuitive appeal.

Zeigler (1976) laid down the first abstract specification of a discrete-event system (DEVS). It consists of a 6-tuple of *‹inputs, states, outputs, $\delta$, $\lambda$, tadv›*, where the sets of input variables, state variables and output variables are defined as for a discrete-time system. A subset of the state variables includes the list of future-event times, as they exist at any time $t_i$. Then the *time-advance* function (*tadv*) is defined as the current time plus the minimum of the future-event times. Thus if the model is in the state $s$ at time $t_i$, it will remain in that state until time

$$t_i + tadv\ (s)\ .$$

The function $\delta$ is the state-transition function, and is split into two parts: one for mapping states to states, and another to be invoked in the presence of external events. Finally, $\lambda$ represents the mapping from the states to the output.

In later chapters, Zeigler (1976) continues this abstract formalism further, specifying a hierarchy of six levels of structural specificity, from mere observance of behaviour to definition of structure. In the hierarchy, Zeigler observes an analogy in that going from the highest level of specification to the lowest is formally equivalent to the simulation of a model by a computer. An alternative simulation hierarchy is described in Chapter 8.

### 2.4.2   Discrete-event simulation paradigms

In discrete-event simulation, we do not stipulate in advance the value of the time increment; on the contrary, this value is determined individually for each time step, based on the component actions of the model. Events are determined by the sequence of each entity starting and finishing activities: the global actions of the simulation are the pursuance of these sequences for several concurrent entity flows. Discretisation of time is thus implicit in the system itself, rather than being explicitly imposed by the simulator. The simulation program is thus concentrated on the events interleaved between activities, and is relatively unconcerned about the periods of active work, leading to a distinctive *inversion* of concern.

The use of the word 'event' here is a specialisation of its everyday meaning, which refers simply to any significant happening. Its etymology is from the Latin *ex-venire* (yet to come). For our purposes we wish to separate the ideas of instantaneous happenings from those with a significant time duration; the former we call *discrete events,* or simply events, and the latter *activities*. This choice enables us to consider time advancing from one event to the next, each event being concerned with a change, for some flowing entity, from one activity to the next. In this way, we avoid an incremental method which would require frequent interrogation of the system state to no avail.

We should also notice that the term 'activity' does not necessarily imply that an entity is performing in any particular way: being gainfully employed or simply waiting for an opportunity for such employment both qualify as activities since they extend over time. Exactly what quantitative limits we use to define an instantaneous event depends on the modeller's view of its significance in the simulated system. There are no hard-and-fast rules. Queueing theory defines customer arrivals as instants, and the same assumption usually carries over to simulation, but we shall meet other circumstances, e.g. a lift system, where the time required for passage through a door is quite significant.

To recap, a discrete-event simulation is a model of a dynamic system which is subject to a series of instantaneous happenings, or events. The events themselves arise out of actions within the simulation carried out at previous events and are used as the basic elements to drive the simulation through a developing sequence of state-changes. The future events which have yet to occur are stored in the future-event set (FES). The advancement of time is accomplished by two primitive operations on the FES: firstly, the set is accessed to provide the next event (i.e. the future event with the minimum occurrence time); secondly, newly derived events may be stored away (*scheduled*) to await future occurrence. Discrete-event simulation is thus a kind of event-driven programming. The algorithmic details of FES scheduling are discussed in Chapter 3.

Zeigler (1976) describes the time advance for discrete-event systems as a means of bridging the gap between one event and the next by assuming nothing happens in the gap. The assumption that nothing happens, or, more precisely, that the state remains piecewise constant between instants, is satisfied by a wide class of real systems. We may call this the *first paradigm* of discrete-event simulation.

A discrete-event simulation consists basically of a parallel flow of entities interacting with resources in activities. The existence of individual flows operating and *interacting* in parallel is the root cause of the difficulties in model-representation (Tocher, 1964). Actually there is no strong reason to continue to distinguish between entities and resources: the latter can be regarded as cyclical entities. Sometimes the term *quasi-parallel* (Kaubisch *et al.*, 1976) is used to describe this mode of simulation, which enables the representation of a parallel system on a sequential computer.

For the paradigm to be successful, every event-notice must thus have a predictable occurrence time when being inserted into the FES. Not all events are directly predictable however, some depending on predictable events of other entities, or maybe some configuration of the model, but the first discrete-event paradigm still can hold up as long as each event activation can be assumed to occur at *some* event. It is non-controversial, being accepted, with more or less recognition, by all discrete-event simulation systems.

We have already noticed that a resource-limited system can be resolved into a statistical half and a structural half. When we say that the structural behaviour of the entity is represented by a computer program, we mean that the advance of the entity through the system is mapped onto the control flow of the program. Complex entity interactions thus correspond to types of control structure which require support from the simulation programming language being used.

The implications of the relationship between entity flow in resource-limited systems and the linguistic expression of program control are being examined in the development of a new simulation programming language called SIMIAN. The control essentials are abstracted from the system and generalised into language features.

The role of the flowing entity can be seen as going through a series of activities in which it is temporarily *engaged* with other entities. The means by which one entity is sought out and put with another into engagement with another is a process of *allocation*. Generally speaking, both the engagement and the allocation phases of development present hurdles to the progress of an entity through the system, but the essential difference between the two is that the engagements consume time, whereas the allocations do not. Consequently, in order to examine the quantitative nature of the bottle-necks in the system, we can do this by surrounding individual engagements with *data-probes*. This more structural simulation paradigm of simulation will be justified during the course of the book. The starting point of the SIMIAN project is the realisation that, in the body of a simulation program text, statements will fall into one of three categories:

  (i)  entities being engaged in activities;
 (ii)  decisions about entity selection and routeing;
(iii)  data-collection statements.

In fact, we could summarise this view in an equation, which we call the *second paradigm of discrete-event simulation*:

$$\text{simulation} = \frac{\text{engagements}}{\text{data-probes}} + \text{allocations}.$$

Thus a simulation program is resolved into three quasi-independent areas: the sequence of engagements, allocations between entities to form engage-ments, and the recording of data from data-probes; the 'division' in the paradigm indicates that the data-probes are concerned solely with the timing aspects of engagements. Without data-probes, there is no output; without allocations, no resource effect; without engagements, no time advance. Thus the control flow of the simulation maps onto three separable domains of the system.

### 2.4.3  Combined discrete-event–continuous simulation

An important distinction can be made between simulations whose inter-actions are continuously effective (usually represented by differential equa-tions), and discrete-event simulation. When the object system is such that it would require both types of time advance to describe it, the first paradigm will not apply. For these situations, we can simulate by attempting to combine discrete-event and continuous approaches. The system can be advanced according to the differential equations up to the most imminent future event, which then occurs, altering the state of the system and allowing the continuous part to take over. Future events are thus used as a sequence of upper bounds on the solution.

    One of the earliest languages to incorporate discrete and continuous simulation into the same package was GASP IV (Pritsker, 1974), extending an earlier discrete-event simulation language, GASP II. Continuous vari-ables, known as state variables, are integrated between adjacent discrete events, so-called 'time-events', according to the simulationist's specifica-tion. Events which occur when the system reaches a particular state of threshold are called 'state-events'. Unlike the time-events they cannot be forecast far into the future and must await the result of integration of the functions which represent the system. However, both types of event are treated similarly in that their occurrence may invoke a subroutine, which may initiate further changes in the system. Cellier (1979) surveys the techniques and methodology involved in combined languages.

In Babich *et al.* (1975) a *significant-event* simulation system is described, incorporating a hybrid time-advance mechanism alternating between clock-pulsed and next-event, which is useful for simulating a system such as road-traffic flow, where the activity in the system changes frequently between periods of slack and complex activity. A clock-pulsed time advance with a fixed time step is used during the periods of complex activity, and during quiescent periods, time is advanced to the next event. A specially denoted event, the so-called significant event, signals the onset of complex activity and the changeover to clock-pulsed time advance. It was developed in the context of simulating switching circuits which are highly synchronised and therefore are usually simulated by a clock-pulsed (i.e. discrete-time) approach.

Hogeweg and Hesper (1979) found useful a form of combined discrete-event and continuous approach to simulation in a biological context, especially for the case where the discrete-event simulation breaks down because of unexpected interrupts. This justification for a combined approach is opposite to the one usually cited (e.g. Cellier, 1979) in which continuous simulation is employed until the system undergoes sudden changes in the model structure or alteration in the settings of parameters.

### 2.4.4  Other discrete-event models

Other OR techniques can be seen as statistical models linked together by interactive structures of some kind. For instance, in the field of project management any project can be broken down into a set of component activities, each with an accompanying begin-event and end-event. Between the activities, various *precedence relations* will hold, to ensure that the proper procedure for the project is performed. For instance, when building a house by conventional methods it is usual to build the walls before putting on the roof. Between any two component activities, either they may take place in parallel, or one must wholly precede the other. For any project for which the duration of each activity is known, or may be estimated, a particular sequence of activities will be critical, in the sense that any delay to these activities will delay completion of the overall project. These activities comprise the *critical path*. Normally the activity times are only estimates of future activity, and may be estimated by assuming a probability distribution.

There is clearly some common ground between precedence networks and discrete-event simulations. For both cases a key concept is that of the *activity*, i.e. an action which takes a significant duration of time to occur, yet the nature of the constraint is different. For precedence nets, the constraint of the commencement of an activity is that all tasks which logically precede it should be completed, e.g. we must ensure that supporting walls are finished before starting the roof. Activity networks are thus statistical models of activity duration with relations of precedence between them, whereas in discrete-event simulation we predicate activity commencement on resource availability, and the sequence of activities is usually regarded as more fixed. While we may identify the evolution of a discrete-event simulation with the progress of individual entities, for a precedence net the corresponding notion would be some kind of global 'project-progress' (Pritsker and Pegden, 1979), and for a project there are unique start and finish events. The prospect of developing common software for both kinds of system is examined further in Chapter 8.

Besides activity networks, there are many systems involving computers and their peripherals linked into networks that come under the heading of *transaction processing* (Moss, 1985), etc. Modern banking systems incorporating electronic funds' transfer are examples of this kind of system. Also models of inventory stock movements, balance of account, and financial models can be seen as discrete-event systems. In principle, all these systems can be simulated by the approaches we shall discuss in later chapters.

Shannon (1975) characterises discrete-event simulation as

> ENTITIES
> having ATTRIBUTES
> interact with ACTIVITIES
> under certain CONDITIONS
> creating EVENTS
> that change the STATE;

which neatly ties up the nomenclature with the basic operations. We might quibble whether entities interact *with* activities — it seems better to think of entities interacting with one another *during* activities. In fact the linking together of interacting entities undergoing the same activity, as an *engagement,* will form the cornerstone of proposals for a new strategy for discrete-event simulation, as we will see in Chapter 8.

# 3

# Scheduling future events

## 3.1 THE PRIORITY QUEUE OF FUTURE-EVENT NOTICES

In the previous chapter we saw that a discrete-event simulation can be typified as a program which is *event-driven*. If the events to which the program is responsive occur in reality, then it would be better classified as a *real-time* program (Young, 1982), as the responsiveness required imposes more severe constraints on the language. Here we are concerned with self-contained programs which generate their own future 'events'. However, there are certain similarities between languages for simulation and for real-time programs, some of which are discussed in Chapter 7.

In a simulation, the 'events' originate from a set of *future-event notices* (FE notices) which is handled by the simulation executive program. An FE notice may be thought of as an entry in a diary denoting a future appointment for, say, a meeting. When the date and hour fall due, the meeting 'occurs'. Often, the activities which take place during the meeting will cause new, future appointments to be 'inserted' into the diary. The 'next event' (appointment) to occur can be found by looking through the diary, in ascending order of time, to the next entry. Occasionally, some entries may need to be postponed or cancelled. Naturally, when appointments have occurred, their entries can be discarded.

Diary operations have direct analogies in the operations of the simulation executive. Instead of 'diary', we call the collection of FE notices the *future-event set* (FES). FE notices are retrieved from the FES in ascending order of time, on which basis the clock time of the model is advanced. Meanwhile, newly created FE notices, resulting from actions performed at event occurrence, may be inserted into the FES (Fig. 3.1). Collectively, the

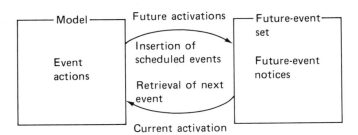

Fig. 3.1 — Relationship between the future-event set (FES) and the activation of the model.

operations on the FES are called future-event *scheduling*.

There are thus two essential pieces of information to be associated with every FE notice:

- the occurrence time of the event,
- a pointer to the actions comprising the event's occurrence.

All data structures for the FE notice will include a reference to the actions of the event. The form of this reference, which could be an address, a label, or a procedure-array index, will often depend on the kind of *simulation strategy* being employed. This is further discussed in later chapters. For the purposes of describing and comparing scheduling algorithms there is no need to continually refer to the existence of the reference, so for the rest of the chapter we shall consider the FES simply as a set of elements, each containing an occurrence time.

Essentially, discrete-event scheduling consists of the following operations:

- the insertion of FE notices into the FES, and
- the extraction of event-notices from the FES in strictly *earliest-out* priority order.

After an event has been activated, its FE notice can be deleted. In some cases, especially when modelling catastrophic events, FE notices are inserted only tentatively, and these may also require deletion before occurrence. The chief problem we address in this chapter is: what is the best algorithm and data structure to perform these scheduling operations?

Scheduling operations can be thought of as processing the elements of a set, where each element has a specific number attached to it (the FE notice's occurrence time), determining its *priority* for removal from the set. Using the nomenclature of Aho *et al.* (1974) the FES is a *priority queue*, i.e. a set whose elements are removed according to their priority. Reeves (1984) suggests 'SIFO' (smallest-in, first-out) as an appropriate acronym. But the FES is not required to be as general as the priority queue as defined by Françon *et al.* (1978), which permits the merging of two priority queues. The FES can be thought of simply as a queue with a priority-defined discipline.

It only remains to decide in what order to remove FE notices, which for some reason have *equal priorities*, for the FES to be a true priority queue. To decide on this question, we should consider the ways in which FE notices can have equal occurrence times. There are three possibilities:

(i) the time base adopted for the program is unable to discriminate between two close occurrences;
(ii) an occurrence is generated to occur at the current simulation time;
(iii) two occurrences have been generated to occur simultaneously, by chance.

Coincident events of type (i) are a headache for simulationists using an integer time base. Since occurrence times usually originate from a random sample taken from a probability density function, the most convenient form for time is obviously a real number. Unfortunately many simulation languages have a long history reaching back to the time when computer memory was in such short supply that integers were preferred over reals for their compactness, and this preference is still supported, for instance in the popular language GPSS. Using languages with an integer time base, the simulation programmer is forced to assume a time-scale factor to reduce the chance of coincidences to what is hopefully a manageable level. Of course, even reals can be equal, but normally their high precision makes it very unlikely for this to be a problem.

It is conceivable that, while processing one event's actions it turns out that another event's actions should be invoked straight away, e.g. when the second event is an immediate consequence of the first. The likelihood of this is determined partially by which simulation strategy is being used, some languages preferring to invoke actions by scheduling events to occur 'now'. By choosing a language with a better simulation strategy we may avoid the problems involved with type (ii) coincidences.

When two events are scheduled to occur at the same time, and neither type (i) nor type (ii) coincidences are involved, then there is nothing that can be done to ensure a faithful ordering of their occurrences. If their actions and effects are independent, then no problem will occur. If not, all we can do is hope that there is no significance in the logical ordering of their actions. In Chapter 8 we put forward some suggestions for dealing with this problem.

If we are forced by an algorithm to make a choice, we may as well choose a FIFO ordering, for want of a better one. Some workers would like to enforce a FIFO ordering in all cases of coincidence, but it can only really be justified in type (iii), and then on rather arbitrary grounds. For type (i) coincidences, the enforcement of a FIFO ordering for time-tied occurrences could lead to serious error in representing the simulation dynamic (Tocher, 1969). For type (ii) coincidences, the occurrences naturally proceed in FIFO order. Notice that a LIFO ordering would be illogical in this case, implying effect preceding cause.

In the early development of simulation languages, the fundamental role of FE scheduling was not always realised. By concentrating on the representation of the entities involved, the occurrence time of events may be treated as just another entity attribute, and in order to advance time, the whole set of currently active entities must be scanned to find the earliest.

Such a treatment of time was frequently encountered in implementations of what we may call the time-cell variant of the *activity-scanning* strategy (Pidd, 1984), which requires that each machine (entity) has its own time parameter (time cell), causing the FES to be scattered among the entity records. But treating time as merely another entity attribute has the effect of denying the application of algorithmic studies of FE scheduling, leading to gross inefficiences especially for large-scale simulations. By treating the set of FE notices as a whole, we can make significant advances in the efficiency of simulation.

For discrete-event simulation, without future events there can be no time advance. While it may be that some problems of priority are influenced by strategy considerations, strategy affects only the form of modularisation describing the model's state changes. So whichever strategy is adopted, future-event scheduling remains the basic algorithmic underpinning of model activation, and the concerns of this chapter can be regarded as fundamental.

The earliest works in this area still feel the need to justify the discrete-event paradigm over incrementation by a fixed amount of time. The first paper to address scheduling problems was the analytical work of Conway *et al.* (1959), followed by further considerations on the nature of time flow in a simulation by Nance (1971). This paper contains some interesting observations, as well as some errors, and stresses the fundamental difference between temporal and state-dependent events. The first paper to bring a detailed and standardised comparison of FES scheduling algorithms to general attention was that by Vaucher and Duval (1975). For additional background to the field of data structures and algorithms, the texts of Aho *et al.* (1983) and Stubbs and Webre (1987) are recommended.

## 3.2   LINEAR-LIST METHODS

### 3.2.1   Simple linear list

Scanning all future occurrence times in the model to select the minimum will certainly be sufficient to perform the function of activating events in the correct temporal sequence. But as the number of FE notices becomes large, this method will become increasingly time-consuming and inefficient. The first improvement, obviating the need to scan *all* future occurrences, is to link together the FE notices in order of their occurrence times.

If we decide to link FE notices in ascending order of occurrence, as in Fig. 3.2, then a simple scan from the low end of the list (i.e. from the earliest

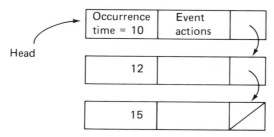

Fig. 3.2 — A singly-linked-list implementation of an FES with three FE notices.

FE notice) will more quickly find the correct insertion position for a new FE notice, by searching for the first FE notice whose occurrence time *exceeds* that of the new FE notice and inserting the new one immediately in front of this notice. To facilitate deletion of a non-current FE notice, corresponding to the cancellation of a future appointment, we would prefer to employ a doubly-linked list, giving the LLF algorithm (Fig. 3.3). (Algorithms are described in Algol 68, a brief summary of which will be found in section 3.6, on page 71).

For the retrieval of the next event, we simply take the element pointed to at the head of the list, and make the following element the new head. On average we would expect the overall scheduling time to be thus reduced by half, at the expense of providing the memory space for the fields containing links.

As with the design of all algorithms, care must be taken to ensure that they are *robust,* i.e. besides performing correctly by delivering all FE notices in earliest-first order, the algorithm will not break down in 'awkward' situations. In the case of the doubly-linked list shown in Fig. 3.4 for example, we should ensure that for any $n$, the FES size, there are $n+1$ possible insertion positions for new FE notices, and that it is quite permissible to insert an FE notice *before* the minimum. Also the algorithm must exit 'gracefully' when a *next-event* call is made on an empty FES, as this is not necessarily an error, e.g. in the case of a combined discrete–continuous model. The seemingly 'incorrect' relationship between the dummy-head FE notice and the first and last elements of the list is to ensure that there is no special-case operations required for insertion in the first or $(n+1)$th positions.

Alternatively we might choose the LLB algorithm, and organise the search of the list structure for the insertion position to be *backwards* from the high end towards the low end. This would entail setting the occurrence time in the dummy header to $-max\ real,$ and changing the inequality of the conditional in the **while** loop to 'greater than'. In practice it is doubtful whether the direction of search of a linear list matters a great deal, although McCormack and Sargent (1981) report consistently better performance in the LLB case. The overall scheduling distribution, which ultimately determines the relationship between input and output sequences of FE notices, may also be significant. We will consider more closely the effect of scheduling distribution on the performance of scheduling algorithms more closely later in this chapter.

### 3.2.2   Indexed linear list

One limitation of the linear list as an implementation of the FES is that there is only one possible starting position for the search. For the LLF algorithm of Fig. 3.3, the starting position is at the 'smallest' (earliest) element, and we have seen how the logic may be changed to start at the largest value of occurrence time (LLB). The most time-consuming step of the scheduling process is the search of the FES to find the insertion point. Clearly, if we can be guided to start searching at a point nearer the insertion point, then we can expect improved performance (Engelbrecht-Wiggans and Maxwell, 1978).

The two-list structure of Blackstone *et al.* (1981) divides the FES into a small, time-ordered list of FE notices near the beginning of the FES, and a larger, unsorted set with occurrence times further out in the future. The cutoff point between the lists is adaptive: it attempts to keep roughly $\sqrt{n}$ notices on the shorter list. The next FE notice is taken from the head of the shorter list, unless it is empty, in which case the cutoff is advanced, and a new short list is selected from the unsorted list.

```
    mode fenotice=struct (ref fenotice pred, succ, real evtime);
    fenotice head:=(head, head, +max real);
    ref fenotice empty=nil;
    real time now:=0.0;
```

\#    The empty list has the header dummy **fenotice**, *head*, prepared to receive the first FE notice. Both *pred* and *succ* pointers refer to *head*. The value +*max real* represents the maximum real value possible in the Algol 68 environment. The first accesses, made during the initialisation of the model, are made through the following procedure:

\#
```
    proc init=(real t) void:
    if pred of head is head        # list empty? #
    then
            pred of head:=succ of head:=
                heap fenotice:=(head, head, t)
    else
            insert (t)
    fi;
```

\#    The first insertion is handled specially, but thereafter the *insert* procedure is used:

\#
```
    proc insert=(real tev) void:
    begin
            real t:=tev+time now;
                        # t is the absolute occurrence time #
            ref fenotice here:=succ of head;
            while evtime of here≤t
            do
                here:=succ of here
            od;
                        # insertion position found #
            pred of here:=succ of pred of here:=
                heap fenotice:=(pred of here, here, t)
    end;
```

\#    Once the initialisation is complete, time is advanced by assigning to the global variable *time now* successive values of occurrence time obtained from the *next event* procedure:

\#
```
    proc next event=real:
    begin
            fenotice current:=succ of head;   #earliest FE notice #
                # remove this FE notice from FES #
            ref fenotice next=succ of succ of head;
            succ of head:=next;
            pred of next:=head;
            succ of current:=empty;
                # deliver the result #
            evtime of current
    end
```

Fig. 3.3 — The LLF scheduling algorithms for the FES implemented as a doubly-linked list. The scanning direction for insertion is *forwards*, from the earlier to the later occurrences.

Initially, the empty list:

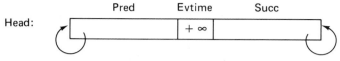

After adding one FE notice with occurrence time $t_1$:

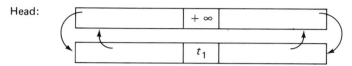

After adding another FE notice, with $t_2 > t_1$:

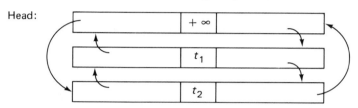

$+ \infty$ denotes the maximum real value in the implementation

Fig. 3.4 — The growth of a doubly-linked list, with header, using the LLF algorithm. The *evtime* field of the dummy *head* element should be set to $- \infty$ if the scanning is to be in decreasing order of occurrence time, and the sense of the conditional in the **while** loop reversed.

Alternatively, we can arrange many possible starting points in an *index*, to be used in a similar way to how a book may be indexed to guide the reader more directly to the part which is of interest. Unlike providing an index set for an indexed-sequential file, we should bear in mind that in a simulation, time is continually advancing, and elements of the FES at the minimum occurrence time are continually being deleted and new elements are being added. Thus the indices will need to be repeatedly updated, which could be a considerable overhead if it is too complicated. For the design of an indexed-linear-list algorithm, there are two main features which must be decided upon:

- how many indices should there be, and
- by what means should they be distributed throughout the set of FE notices?

An interesting and very simple method of indexing is to maintain a single median pointer, as an alternative starting position for the search, i.e. to divide the FES into two equally sized sets (Davey and Vaucher, 1980; O'Keefe, 1985). From the practical point of view, this is a straightforward procedure, requiring only trivial updating, without the overheads of a complex set of indices, yet can easily be made self-optimising.

The next category of indexed methods is characterised by having many indices equally spaced in time (see Fig. 3.5). One method, described by

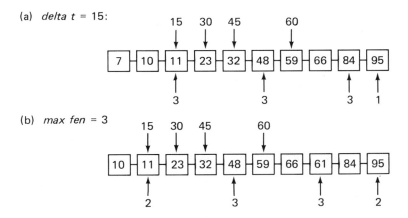

(a)  *delta t* = 15:

(b)  *max fen* = 3

(c)  After extraction of min (7) and insertion of (61)

(a) The upper pointers represent an indexation of the FES with *delta* $t$=15, the lower pointers an indexation with *max fen*=3.
(b) The same FES after extraction of the minimum FE notice (7) and insertion of a new one at (61).

Fig. 3.5 — An indexed linear list, with *delta t* and *max fen* indexations.

Vaucher and Duval (1975), uses a constant, *delta t,* to separate dummy entries in the FES which are used as the index set. When a dummy is selected as the next event, a new dummy is created. Although this will incur an overhead in processing time, it should be outweighed by the benefit of shorter searching time, as long as an effective value of *delta t* can be gauged. If the value is too small, there are too many indices and the overhead is high. If too large, the indexation is less effective.

Moreover, the number and distribution of FE notices within the FES may change during the simulation run, and an effective value of *delta t* would seem to depend on the characteristics of the particular simulation model being executed. It seems unlikely that one *delta t* value would serve adequately in all situations. Nevalainen and Teuhola (1979) investigate a method in which feedback is used to adjust the value of *delta t*. They claim a method which is as effective as that of Vaucher and Duval's, as long as the changes to the FES are not too radical, but which does not require the modeller to provide an estimate of *delta t*.

As an alternative to having dummy FE notices equally spaced in time, they may instead be separated by equal numbers of FE notices. During insertion, a count can be made of the number of FE notices scanned. If this exceeds a certain maximum, *max fen,* then a new dummy FE notice is inserted in the list. In this way the maximum number of scans is controlled and more efficient access can be gained to sections of the list where FE notices are relatively densely distributed (see Fig. 3.5). The indexation is thus adaptive to the distribution of FE notices (Wyman, 1975, Method C). The median-pointer method of Davey and Vaucher (1980) can be thought of as equivalent to maintaining a *max fen* value at around $n/2$. A combination of adaptive and *delta t* methods has also been found effective (Comfort, 1979).

As more refinements are added, generally better performance is achieved, and reduced sensitivity to the distribution of FE notices is obtained. Henriksen (1977) extends the indexed list of Vaucher and Duval (1975) and considers using a binary search of the index set (in vector form) to find the starting position for the search. Henriksen's control of the sublist size has been called 'a beautiful compromise between cost and effectiveness' (Kingston, 1984). As a sublist is searched during an insertion, a count is kept of the number of FE notices scanned. When this exceeds a certain maximum (usually fixed at four), a vector element is set to point to the FE notice currently being scanned, and the count reset to zero. Rarely does this necessitate an increase in the vector size.

The method has been refined (Henriksen, 1983), using a binary tree to find the starting position of the search. The *max fen* value of 4, which is used in both methods, is claimed to be optimal, and the latter method has been incorporated into the GPSS/H simulation package (Henriksen and Crain, 1982).

Without going as far as having a binary tree to select an index, the next stage of sophistication comes with an extra level of indexing. A set of secondary keys indexes the index set. The addition of an extra level of indexing reduces the complexity of the algorithm from $O(n)$ to $O(\sqrt{n})$ (Wyman, 1975), assuming that the FE notice density is uniform over the whole range, which should be true in a dynamic sense, if the simulation has reached a steady state. In any case, some improvement should be gained.

This method can be further enhanced with yet another level of indexation, and so on, until the eventual structure is like a tree structure with FE notices on the leaf nodes, broadly resembling that of Henriksen (1983). Such a kind of multi-list structure is suggested by Comfort (1979). In addition to allowing searches at the FE notice level to cause creation of keys in the next higher level, a search at any level may cause higher-level keys to be created, even creating that level if necessary.

A two-level indexed list was suggested by Franta and Maly (1977, 1978), aiming to combine *delta t* and *max fen* methods. It contains three components:

  (i)  a linked list of FE notices divided into sublists;
 (ii)  a linked list of primary and secondary keys;
(iii)  an array indicating which keys are primary.

The primary keys point to dummy notices, as in the *delta t* method, whereas the secondary keys partition the set into sublists containing approximately equal numbers of FE notices. The structure is adaptive insofar as when a sublist, i.e. a linked section of the FES indexed by a single dummy-event notice, exceeds a certain value of *max fen*, FE notices are moved into adjacent sublists with consequent alteration of index values.

The main shortcoming of using indexed linear lists for FES operations is their lack of flexibility in response to changes in the distribution of the FE notices. Attempts to optimise the list parameters, be they number of indices, *delta t, max fen,* or number of index levels, always assume a constant flux of FE notices of known distribution. We will return in section 3.5 to consider what can justifiably be assumed about a typical simulation.

## 3.3   HEAPS

The *heap* (Williams, 1964) is a data structure which, we may presume, was invented for the very purpose of performing the operations of an FES in discrete-event simulation, because around the same time as the abstract heap algorithm was announced in the *Communications of the ACM*, the same author was working on the Algol-based simulation language, ESP (Williams, 1963).

A heap is a set of elements between which there is a father-and-son relationship such that the *heap property* holds, i.e. a father element is less than or equal to its sons. The heap is a type of balanced-tree structure which preserves the heap property during additions and deletions.

The heap can be arranged as an array *a* with elements indexed from 1 to *n*. This is the common form of its implementation, called (Jones, 1986) an *implicit* heap. The father–son relationship becomes a simple calculation on the array indices: assuming each father has two sons, the heap property can be defined as

$$a[i] \leq a[j] \text{ for } 2 \leq j \leq n, \ i = j \div 2 \ .$$

With this arrangement, for any element $a[k]$ considered as a father, its two sons are $a[2k]$ and $a[2k+1]$. Conversely, for any element $a[k]$ considered as a son, its father is $a[k \div 2]$. Element $a[1]$ is thus always the least member of the set. Insertion and deletion operations using the heap will have $O(\log n)$ complexity (Williams, 1964).

Elements of the heap migrate upwards or downwards in 'sifting' operations. Essentially, sifting involves a series of exchanges between father and son in order to satisfy the heap property. As the shape of the heap solely depends on the *number* of items it carries (Fig. 3.6), the structure does not require any balancing operations to be performed.

The smallest element, corresponding to the earliest FE notice, resides at the root. It may be extracted by first making a copy of it, and then overwriting it by an element with no sons, most conveniently the element residing at *heap[up]*, the base of the heap. The heap property is then

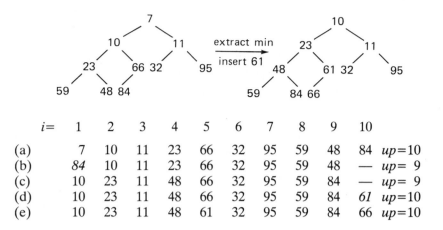

|  | $i=$ | 1 | 2 | 3 | 4 | 5 | 6 | 7 | 8 | 9 | 10 |
|---|---|---|---|---|---|---|---|---|---|---|---|
| (a) | | 7 | 10 | 11 | 23 | 66 | 32 | 95 | 59 | 48 | 84 *up*=10 |
| (b) | | *84* | 10 | 11 | 23 | 66 | 32 | 95 | 59 | 48 | — *up*= 9 |
| (c) | | 10 | 23 | 11 | 48 | 66 | 32 | 95 | 59 | 84 | — *up*= 9 |
| (d) | | 10 | 23 | 11 | 48 | 66 | 32 | 95 | 59 | 84 | *61* *up*=10 |
| (e) | | 10 | 23 | 11 | 48 | 61 | 32 | 95 | 59 | 84 | 66 *up*=10 |

(a) The initial configuration of the heap. Variable *up* contains the number of items.
(b) The minimum element (7) is extracted, and overwritten with the base element, i.e. *heap* [*up*], containing (84).
(c) To restore the heap property, (84) is sifted down until it is greater than its father element.
(d) A new element (61) is added. It occupies the base element position, initially.
(e) To restore the heap property, (61) is sifted up until it is greater than its father element.

Fig. 3.6 — The binary heap as an FES. The initial set of FE notices is here initialised by FE notices with occurrence times in the order {59,32,7,48,66,11,95,10,23,84}. The implicit representation is in the form of an array, whereas the same arrangement can be represented explicitly as a tree graph. Notice that the implicit format corresponds to a top-down, left-to-right reading of the explicit format.

restored by swapping this element with the smaller of its sons, whose place is taken in turn by the smaller of its sons, and so on, producing a ripple of exchanges down through the structure.

Insertion may be performed by placing a new element at the base of the heap and then restoring the heap property upwards towards the root. These operations are called *sift down* and *sift up*, respectively, and are detailed in Fig. 3.7. Deletion can be handled by overwriting the offending element by the smaller of its sons, and restoring the heap property by a series of *sift down* calls.

Besides the implicit heap structure (HPA), in which index offsets for referring to father and son elements are calculated, we may use pointers to connect fathers with sons, to avoid the overhead of calculating indices, but at the expense of using more memory per element.

The initial appeal of the heap method lies in the ready availability of the next event, which is at its root. This has led to its recommendation for FES

```
int up:=0;  #number of elements on heap #
[1:500] real heap;  # maximum of 500 assumed #
proc init=(real t) void:
    (up plusab 1;
    heap [up]:=t;
    sift up (up) );
proc swap=(int i,j) void:  # interchange two heap items #
    (real h=heap [i];
    heap [i]:=heap [j];
    heap [j]:=h);
proc sift down=(int i) void:
while  # ensure that fathers≤sons #
    int j:=2*i; int k:=j+1;  # sons' indices #
    int m;
    if k≤up  # prevent index being outside bounds #
    then m:=if heap [j]<heap [k]
                then j else k fi;  # m is index of min son #
        true
    else
        false
    fi
do
    if heap [i]≥heap [m]  # father≥min son? #
    then
        swap (i,m);
        sift down (m)
    fi
od;
proc sift up=(int i) void:
while
    int j:=i over 2;  # father's index #
    if i>1
    then
        heap [i]<heap [j]  # i.e. son<father? #
    else
        false
    fi
do
    swap (i,j);
    sift up (j)
od;
```

Fig. 3.7 — HPA operations on the heap to perform FES scheduling.

operations by Knuth (1973a) and Gonnet (1976). One disadvantage for use as an FES is that it requires the size of the event set to be specified in advance. Another disadvantage is that the heap is 'unstable' in that it does not preserve the FIFO ordering for FE notices with equal occurrence times. This can be overcome by the incorporation of an extra field indicating the time at which the FE notice was scheduled (Comfort, 1979).

There are several variants of the heap structure which have been proposed for FES operations. The 'fertility' of the heap, i.e. the number of sons attached to each father, can be increased to three, to give a *ternary heap* (Reeves, 1984), Fig. 3.8. The sifting algorithms can easily be accommodated to this change.

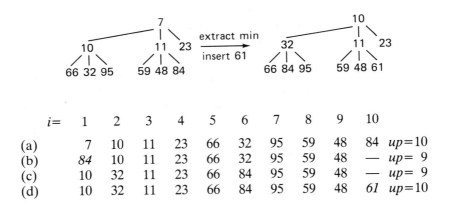

| $i=$ | 1 | 2 | 3 | 4 | 5 | 6 | 7 | 8 | 9 | 10 | |
|---|---|---|---|---|---|---|---|---|---|---|---|
| (a) | 7 | 10 | 11 | 23 | 66 | 32 | 95 | 59 | 48 | 84 | *up*=10 |
| (b) | *84* | 10 | 11 | 23 | 66 | 32 | 95 | 59 | 48 | — | *up*= 9 |
| (c) | 10 | 32 | 11 | 23 | 66 | 84 | 95 | 59 | 48 | — | *up*= 9 |
| (d) | 10 | 32 | 11 | 23 | 66 | 84 | 95 | 59 | 48 | *61* | *up*=10 |

(a)–(c)  The initial stages are similar to those of Fig. 3.7, except that the sons of element with index $k$ have indices: $3k-1$, $3k$ and $3k+1$.

(d)  A new element (61) is added. It occupies *heap* [10], the base element position. In this case, it is already greater than its father, at *heap* [3], so *sift up* will have no effect.

Fig. 3.8 — Implicit and explicit representations of the ternary heap.

There are several other heap-based data structures that have been considered for an FES:

- *leftist tree* (Crane, 1980; Knuth, 1973b);
- *pagoda* (Françon *et al.*, 1978);
- *skew heap,* (Sleator and Tarjan, 1983, 1986). — note that the published algorithm does not handle elements with equal priority correctly (Jones, 1987);
- *binomial queue* (Brown, 1978);
- *pairing heap* (Fredman *et al.*, 1986; Stasko and Scott Vitter, 1987).

## 3.4   BINARY SEARCH TREES

### 3.4.1   Basic insertion and extraction algorithms

For a sorted set of elements to be stored in a *binary search tree*, one item is held per internal node, arranged in the following order: if node *x* contains an item with key *k*, then every item in the left subtree has key less than *k*, and every item in the right subtree of *x* has key greater than *k*. Compared to the heap, items in a binary search tree are sorted 'horizontally' rather than 'vertically'.

Following Evans (1983), using a binary search tree to contain the FES would require the data-structure declarations, initialisations and procedures of Fig. 3.9.

```
mode node =struct (ref node llink, rlink, real evtime) ;
ref node empty=nil;
ref ref node root:=heap ref node:=empty;

proc insert=(real fut, ref ref node node) void:
if node is empty
then node:=heap node:=( empty, empty, fut )
else insert (fut, if fut≤evtime of node
                 then llink of node
                 else rlink of node
                 fi)
fi;

    # oscan retrieves all FE notices in one invocation #

proc oscan=(ref ref node node) void:
while node isnt empty
do
    oscan (llink of node);
    clock time:=evtime of node;
        # carry out state-changes;
            generate more future events,
            inserting them into the future-event set #
    node:=rlink of node
od
```

Fig. 3.9 — The binary search tree as FES, using the OVS method. N.B. The use of the term **heap** here denotes the use of dynamic memory allocation. There is no connection with the heap algorithm of the previous section

New items are *insert*ed by finding a suitable leaf node and creating a new node in its place. An FE notice scheduled to occur five time units hence may be stored by

```
insert (5.0+now, root);
```

where *now* is a variable containing the current value of simulated time. Each FE notice has its predecessors in its left subtree and successors in its right. In addition, FE notices with equal occurrence times will be retrieved in FIFO order.

Deletion of an arbitrary element is a little awkward, however. The difficulty arises when the element to be deleted also resides on the invocation stack, set up to enable the recursive call of *oscan*. Clearly any such element deleted from the tree must also be removed from the stack, which is normally inaccessible to the programmer. In any practical implementation the invocation stack must therefore be provided explicitly by the programmer. Alternatively, the rather unclear code of Kingston (1984) may be followed.

A single left-to-right in-order traversal of the tree, carried out by a call of the *oscan* procedure, will drive the simulation through its whole sequence of state-changes. The traversal path will visit each FE notice in ascending order of time, and will include new FE notices that have been inserted in the meantime. Whenever the traversal passes to the right subtree of a node, the current node, referring now to a past event, is deleted by overwriting its reference with that of its right subtree. This 'overlapping' traversal (OVS) has the effect of clearing out all the old event nodes (see Fig. 3.10).

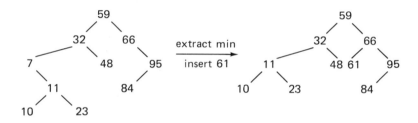

Fig. 3.10 — Binary search tree as FES. The minimum element lies at the leftmost node. After extraction, its reference is overwritten by its *rlink*, thus promoting its right subtree. Continuation of the *oscan* invocation finds the new minimum of (10). The insertion of a new FE notice at (61) will be included in the *oscan* traversal.

Without an overlapping traversal, the root's FE notice, arbitrarily determined by the initial event encountered in the model, would remain undeleted. After the left side of the tree had disappeared, the active part of the tree would be strung out like a kite on an ever-lengthening string of *rlinks* containing obsolete events.

A similar tree-traversal mechanism has been suggested (Knuth, 1974), but not in a simulation context. Kingston (1984) uses a similar algorithm, but without a recursive traversal and using explicit 'backlink' pointers to the parent of each node.

Binary search trees possess features which are intrinsically advantageous for use as an FES, such as $O(\log n)$ insertion complexity, but they can be rather sensitive to the input order of FE notices. For instance, Vaucher and

Duval (1975) found that if the scheduling distribution is such as to lead to FIFO processing of the FES, i.e. if the new FE notice to be inserted is always greater than the maximum, then the tree performs badly compared with a linear list.

To counter this problem, Jonassen and Dahl (1975) put forward the *p-tree,* to combine the advantages of linear lists with the efficiency of tree structures. A p-tree consists of a linear list of items sorted in decreasing order, called the 'left path', such that connected to each node of the left path (except the last) is a (possibly empty) p-tree, called the 'right subtree', whose nodes contain values between $x$, the associated node on the left path, and the left successor of $x$. The terminal node on the left path is the element with smallest key value. Jonassen and Dahl (1975) showed that on average the performance of the p-tree was $O((\log n)^2)$.

### 3.4.2  Balanced trees

The advantage of trees lies in the logarithmic relationship of the number of comparisons required to schedule an FE notice to the number of items stored. However, trees possess two disadvantages: in inserting, each item encountered in searching requires two comparisons, one to check if the item exists, then another to test its value; also, the logarithmic relationship may be lost if the tree is allowed to get too far out of balance. For a tree with all nodes in a linear sequence, as would occur with a strictly FIFO processing of the FES, the efficiency could be considerably worse than a list. Trees which control their balance seem to offer one solution. We can see that with our overlapping scan (OVS) the root is kept within the bounds of the set, but this crude balance is obtained somewhat haphazardly and may perhaps be improved upon.

The AVL tree (Adel'son-Vel'skii and Landis, 1962) uses a mechanism to prevent the *heights* of both subtrees of each node from differing by more than one. The height of a subtree is defined as the length of the longest path from the root to a leaf node. For each node the current balance factor of $-1$, $0$, or $+1$ is stored. If an insertion or deletion threatens to cause the balance of any node on the access path to go beyond these limits, the subtree must be rebalanced. Balancing is performed by applying either single or double *rotation* (Fig. 3.11), which is accomplished simply by a cyclical exchange of pointer values. The complete algorithm is given in Wirth (1976).

For FES use, an overlapping scan is applied to the AVL tree, so that the most recently visited node can be deleted (Fig. 3.12). However, the shape of the tree being under rigid control prevents a complete scan in one invocation. For every access, a complete cycle down from the root and back again, rebalancing *en route* where necessary, is obligatory.

### 3.4.3  Partial balancing

One problem with balancing a tree used for an FES is that the range of key elements is continually changing owing to time advance. The small elements

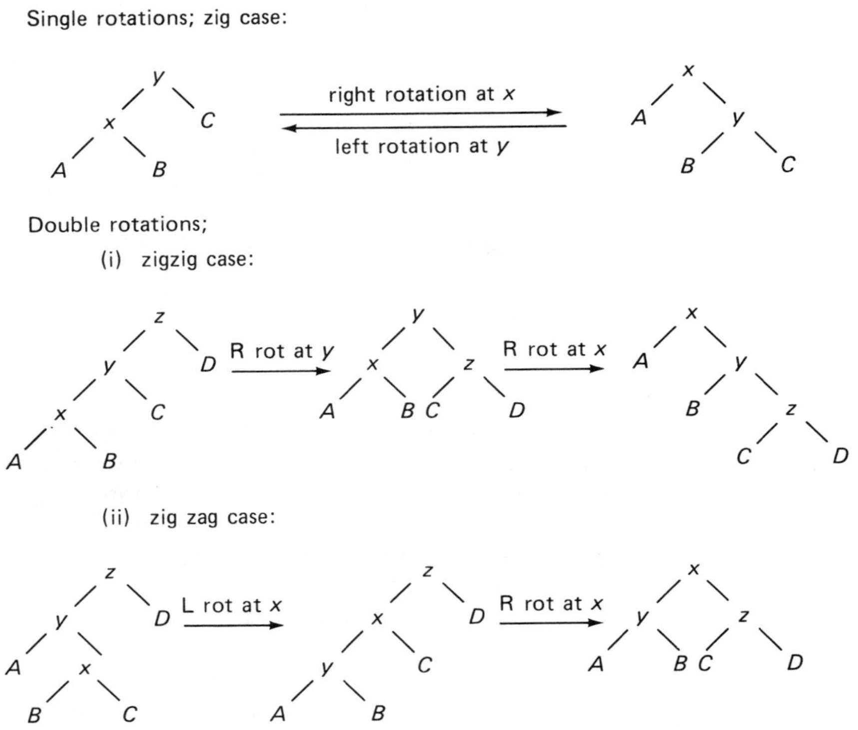

Single rotations; zig case:

Double rotations;
    (i)  zigzig case:

    (ii)  zig zag case:

Fig. 3.11 — Rotations of a binary search tree. *A*, *B*, *C* and *D* represent arbitrary subtrees. Notice that for all rotations, the horizontal sequence of nodes is preserved.

get deleted and the key value of insertions increases. Thus any static indexing or tree balancing will need to receive continual updating. The AVL tree would possibly be able to adapt to a well-balanced shape with just a few rotations if this were not so, but for FES scheduling, frequent rebalancing seems necessary. Perhaps, as Foster (1973) suggests, we require a method of balancing which is less strict.

The *splay tree*, put forward by Tarjan (1983, 1987) and Sleator and Tarjan (1985), uses a restructuring heuristic called *splaying* which moves a node towards the root by performing a sequence of rotations. Instead of affixing balance information to each node, the insertion routine takes note of the local shape of the branch of the tree which is currently being traversed. There are three cases to consider: zig, if the parent is the root; zigzig, if the parent and grandparent are both left or both right sons; and zigzag, if the parent and grandparent are different types of sons. For each case, the

(a)

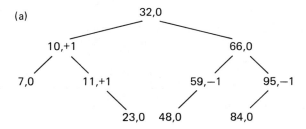

after extraction of the minimum FE notice, and insertion of (61):

(b)

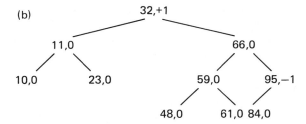

after two more extractions and insertions:

(c)

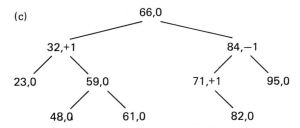

Fig. 3.12 — AVL tree as FES. Balance factors are given alongside the node key. (a) Initial tree. For the extraction of the minimum and insertion of a (61) element, no restructuring is called for (b). However, if we take the FES through two more occurrences, at 10 and 11, inserting FE notices at (71) and (82) respectively, we arrive at the completely restructured AVL tree (c).

appropriate rotation is applied, willy-nilly. The overall result of these reorganisations is to advance nodes on frequently accessed branches towards the root, making them more readily available. The shape of the tree thus adjusts itself to the pattern of access. Alternative implementations are discussed by Mäkinen (1986).

For an unbalanced tree, the normal progress of a simulation will have the effect of increasing the size of the right subtree at the expense of the left. In Evans (1983, 1986b), partial-balance mechanisms are proposed which monitor, in different ways, the relative sizes of these subtrees and employ a criterion by which to trigger one or more left rotations around the root, thereby hopefully gaining a dynamic balance in the long run. The new root must also be introduced at the bottom of *oscan*'s invocation stack. We let the balance of the root's subtrees take care of itself while ensuring that the root more or less keeps up with the median FE notice.

The first method (LTE) rebalances by a series of left rotations until half or more of the nodes lie on the left of the root. This is invoked only when the left subtree of the root becomes empty. Although a somewhat sluggish method, we would expect the left subtree to contain around 25% of the nodes in the long run.

The second method (CLR), which makes use of a node-count of both left and right subtrees, is more attentive to balance by not waiting until the left subtree has completely vanished. Instead we trigger a single left rotation when the right tree population exceeds five-eighths of the total:

$$R > (L+R)(5/8)$$

where $R$ and $L$ are the right and left subtree sizes, respectively. This is not an arbitrary trigger, but aims to maintain 50% of the nodes on each side by transferring one sub-subtree, with expected size one quarter of the total, from right to left.

The third method (STL) monitors instead the length of *oscan*'s invocation stack, which for a balanced tree should be approximately the logarithm to base 2 of the tree size. A single left rotation is performed when the stack length decreases to two less than this figure.

## 3.5  COMPARISON OF METHODS

### 3.5.1  The hold model

If we are going to decide between the various structures and algorithms to find the best to support FES operations, then we have to estimate in some way what is the typical kind of work to which the method will be subjected. Two essential operations are those of extraction of the next FE notice, and insertion of a new FE notice, so clearly any implementation of the FES is going to involve some searching, whether for the next FE notice, or for the insertion position for a new one. The efficiency of the operations will also be affected by the statistical properties of the sequence of generated FE notices.

The first serious investigators of this problem (Vaucher and Duval, 1975) suggested the *hold model* in which the typical scheduling action is judged to be: extracting the minimum FE notice, advancing the model clock to its occurrence time, and scheduling a single FE notice, obtained by randomly sampling from a given scheduling distribution. The name is taken from the *hold* operation of the simulation language Simula (Birtwistle *et al.*, 1973). Since it was first put forward, the hold model has become widely accepted as a standard benchmark in empirical comparisons.

As with all models, the hold model is an approximation to the real situation. Let us examine its assumptions. Firstly, only one FE notice is scheduled for each event activation. While this may be typical, there are plenty of situations, even in simple simulations, where this is not the case. For instance, in simulating arrivals (nearly all simulations generate the arrival of entities into the model), there are often two FE notices to be scheduled in the actions of the arrival, one for the next activity of the arrived entity, and the other to generate the next arrival.

Having only one FE notice to schedule for each next event extracted also means that the size of the FES remains constant. While this may be convenient for testing purposes, it is unlikely to occur in realistic simulations. In general it is difficult to put any prior bounds on the size of the FES.

In the hold model, insertion and extraction of FE notices occur alternately. But this is clearly not the general case. Any scheduling method which depended on the certainty of alternating insertion with extraction would be 'cheating' the model. In fact there is a variation on the binary heap method (the modified heap of McCormack and Sargent, 1981) which uses the next FE notice for insertion as the base element for the overwriting of the current minimum. Although it performs very efficiently, the method cannot be regarded as practical.

Some studies have urged that methods should be 'stable' with respect to FE notices with equal occurrence times, and take pains to emphasise that coincident FE notices should be extracted in FIFO order, i.e. in the order in which they were originally entered into the FES. As we have mentioned before in this chapter, this difficulty can be eased by adopting a real-number time base. A better solution would be to deliver all coincident events together and then use some intelligent means to arbitrate between them. In an interactive environment the procedure may be assisted by the intervention of the user. We will return to these ideas in Chapter 8.

The most important assumption made in the hold model is the type of scheduling distribution with which new FE notices are scheduled. Most studies choose distributions from those suggested by Vaucher and Duval (1975). They are:

(1)  exponential, with mean 1.0;

(2)  uniform over [0,2];

(3)  uniform over [0.9 to 1.1];

(4)  bimodal, with 90% uniform over [0,0.095] and 10% uniform over [9.5,9.595];

(5)  constant value of 1;

(6)  from {1,2,3} with equal probabilities.

While these distributions form a set of widely different characteristics, it cannot be said that it is at all typical for a simulation to have *all* of its FE notices determined by only one scheduling distribution (McCormack and Sargent, 1981).

The methodology of the hold model, as advocated by Vaucher and Duval (1975), has been fairly closely followed in subsequent studies. Basically, the time taken to perform a hold operation, consisting of one extraction of the minimum FE notice followed by a single insertion of a new FE notice, under fixed conditions of scheduling distribution and FES size, is observed. To eliminate spurious effects, several thousand holds are repeated and their times averaged. Each measurement can be repeated using the antithetic conjugate of the original random-number stream, to reduce sampling error.

Besides the scheduling distribution under test, the initial distribution of FE notices will also have an effect on a method's performance. We need to avoid any bias deriving from the initial distribution, by starting with a set which approximates the steady-state distribution of FE notices. But the steady-state FE distribution is also determined by the scheduling distribution, which prevents us from using the same set of initial occurrence times for comparisons between different scheduling distributions.

In Vaucher and Duval (1975), each structure was initialised with the same set of FE notices for each condition and then subjected to two holds from the scheduling distribution under test. Other investigators have followed suit, the most ingenious being Jones (1986) who 'grew' an FES from sequences of FE insertion and extraction, with the probability of insertion slightly higher, halting the process when the FES had reached the required size.

There remain some effects for which it is difficult to control. No doubt the hardware architecture and operating system will have an effect. Garbage collection would have a considerable effect if it was allowed to proceed during the timings. Of course it is not entirely fair to leave garbage collection out of the picture, as it is certainly implicated in the provision of new space in the dynamic storage area, but it is too dependent on local conditions to have any general significance.

One way of avoiding the criticisms of the hold model is to concentrate on particular models which are deemed to be representative of a larger class of simulations (McCormack and Sargent, 1979). But for this approach it is difficult to avoid the accusation of arguing from special cases. Also what conclusions can be drawn when results from one type of experiment conflict with those from another?

The main comparisons we shall review here are those of Vaucher and Duval (1975), Comfort (1979), McCormack and Sargent (1981), and Jones (1986).

### 3.5.2   Empirical findings
Since the essence of FES operations is a search through a set of elements, let us consider, for each broad class of structure, the information gained by a single comparison on the remainder of the set. For a linear list, a comparison gives only information about one element, i.e. the size of the set remaining to be searched is reduced by only one. For a binary search tree, the remaining set is reduced by approximately half, assuming the tree is

balanced. In the case of the heap, the situation is intermediate, as the elements are not totally ordered with respect to one another. On this basis, it seems that a binary search tree should be the best structure, at least asymptotically, if only there could be a way of ensuring the tree's balance.

From these initial observations, it is clear (Knuth, 1973b) that the number of elements in the set is an important condition on the behaviour of different structures: if the number is small, then a linear list will be effective; otherwise we will need to employ an $O(\log n)$ method (Nevalainen and Teuhola, 1979).

The comparisons of trees and lists in Vaucher and Duval (1975) come down strongly in favour of indexed linear lists, and the authors recommend that more work be carried out on finding an empirical formula to make them more adaptable. The tree methods considered are the p-tree (called the 'end-order' tree) and a complicated version of binary search tree ('post-order') with provision for a linear list at each node to contain a stack of coincident FE notices. Heaps are rejected because of non-FIFO retrieval of coincident FE notices. The poor showing of the post-order tree on distribution number (3) caused it to be rejected as a candidate for FES operations.

This finding set the scene for later research, and the concentration on indexed linear lists. Comfort (1979) compares five varieties of indexed linear list and two types of heap. Two *max fen* indexation methods, AL ↑ k and I/AL, were found to have $O(1)$ behaviour. The author gives details of the optimisation of the indexation parameters. Since the AL ↑ k method has an unpredictable demand on memory, being able to dynamically adjust the number of index levels, an I/AL, with a *max fen* between 4 and 6, was recommended.

McCormack and Sargent's (1981) survey took into consideration two properties of the scheduling distribution, the coefficient of variation and %$F$, a measure of the average rank position of an FE-notice insertion. This information was made use of in determining scan directions. Whether or not it is justifiable to make use of such information is arguable. They found the best methods, from a set including different kinds of indexed linear lists and heaps, to be the binary search indexed list proposed by Henriksen (1977) and the modified heap. These methods are also less sensitive to different scheduling distributions.

Investigations of FES methods received a new impetus with the work of Kingston (1985) on tree algorithms, and, Jones (1986) who considers priority queue structures that had not previously been investigated under the hold model. Three groups of algorithms are investigated. The linear list, implicit heap and leftist tree are the so-called 'classical implementations'; the two-list of Blackstone *et al.* (1981) and the indexed linear list of Henriksen (1977) are 'special-purpose implementations', and thirdly the 'nearly optimal' implementations: binomial queues, pagodas, skew heaps, splay trees and pairing heaps, a group of algorithms associated with theoreticians Robert Tarjan or Jean Vuillemin.

Jones compares methods under five continuous scheduling distributions. The measurements were taken over 1000 replications for 28 FES sizes, from

1 up to around 11,000 FE notices. So it seems that for the largest FES sizes, around 90% of the elements are unaffected by the *hold* repetitions under investigation. Thus the closeness of the initial FES to the shape of the expected steady-state FES assumes great importance.

The conclusions are, as is normal with most hold-model comparisons, fairly equivocal. No one algorithm performed best over all distributions and range of future-event set sizes. One may say there are two competitors for the best overall performance: the splay tree and Henriksen's indexed list, with the binomial queue a good third. However, some uncertainties about these experiments remain. Too little discussion is offered on the empirical results. Compared with other published graphs, they seem to fluctuate abnormally, with frequent departures from monotonicity, despite the attention paid to detecting perturbations.

To sum up, we may consider the three main groups of algorithms in turn. For linear lists, in order for them to have reasonable performance for large ($\sim$100) FES sizes, some form of indexation seems necessary. The problem with indexed linear lists is that they usually require tuning to deliver their best performance, as they are often sensitive to the FE distribution. From a practical point of view it is not easy to estimate the distribution of future events, which places methods based on indexing at a disadvantage. The claimed efficiency of the two-level list of Franta and Maly (1977) was not substantiated when tested by McCormack and Sargent (1981).

Henriksen's (1977) algorithm performs well in most tests under the hold model, but performs less well in a test based on particular models (McCormack and Sargent, 1979). Kingston (1984) shows that Henriksen's algorithm has an average cost, in terms of number of comparisons, which is independent of the distribution of insertions.

From the experimental results, it appears that the efficiency of heaps is disappointing. The leftist-tree was shown (Jonassen and Dahl, 1975) to be inferior to the AVL tree for large FES sizes. The modified heap produces good results with the hold model, but is impractical, relying on a strict alternation of extraction of minimum and insertion operations. However, the ternary heap (Reeves, 1984) was found to be competitive with an optimised variant of Wyman's (1975) adaptive linear list, with a *max fen* indexation.

Relatively little attention has been paid to the use of trees for FES operations (McCormack, 1979) compared with the interest in variations on an indexed linear list. Myhrhaug (1965), as reported in Vaucher and Duval (1975), found that tree algorithms out-performed the simple linear list for large event sets, and for distributions with large variance. He also found the balanced tree worse than the unbalanced tree, only becoming better for event sets larger than 400. The algorithms were not described in detail.

The binary search tree's vulnerability to FIFO scheduling of future events, which produces a degenerate tree, was behind the design of the p-tree (Jonassen and Dahl, 1975) which aimed to combine the advantages of both tree and linear list. The authors show that it is out-performed by the AVL tree for a set of more than 200 events.

Performance comparisons (Evans, 1986b) using the hold model between the AVL tree, heap mechanisms (HPA and HPB), and a simple linear list searched in descending order (LLB) are shown in Fig. 3.13. The HPB method uses the modified heap of McCormack and Sargent (1981) and explicit pointers from father to sons. The poor performance of these methods against OVS leads us to doubt whether the effort expended in balancing is worthwhile. Foster (1973) comments that strict balancing of trees is only valuable when the data are fairly static, which is not the case in simulation.

The results for partially balanced trees (Evans, 1986b) are given in Fig. 3.14. For both CRL and STL methods, the standard parameters, namely 0.625 and 2, were allowed to vary within a small range. In neither case did the variation make a significant difference to the results using the standard values. Again the results with just an overlapping scan (OVS) seem better. It seems that for FE operations, even partial balancing is not worthwhile.

Comparisons of tree methods against indexed linear lists normally indicate the superiority of the latter. For instance, Nevalainen and Teuhola (1978) found an indexed-linear-list method slightly better, whereas Kingston (1984) warmly recommends Henriksen's (1977) algorithm over the binary search tree and the p-tree.

### 3.5.3   Theoretical studies

There are basically two approaches open in the evaluation of algorithms: theoretical and empirical. So far we have looked at findings from the empirical approach. Theoretical studies aim to obtain algorithms which are intrinsically good without regard to particular architectures, operating conditions or the programming language involved. FE scheduling algorithms are usually classified as to their run-time efficiency (so-called algorithmic complexity) with respect to the size of the FES ($n$). Thus linear lists are $O(n)$ methods and trees are $O(\log n)$ methods. In addition, worst-case analyses are traditional, although this point has been disputed recently, especially in the case of self-adjusting data structures such as splay trees (Tarjan, 1987), where the appropriate measure of complexity should be *amortised* over many executions.

By amortisation is meant (Tarjan, 1985) averaging the running time of an algorithm over a worst-case sequence of executions. This complexity measure is meaningful if successive executions of the algorithm have correlated behaviour, as occurs often when manipulating data structures, and especially in FES operations. Amortised complexity contains aspects of worst-case and average-case analysis, and for many problems provides a measure of algorithmic efficiency that is more robust than average-case analysis and more realistic than worst-case. The amortised complexity of Henriksen's algorithm is derived in Kingston (1986).

Various theoretical studies of algorithms for priority queues have been carried out, but theory can only act as a guide because the results are less discriminating, and cannot identify particular usages so specifically. For

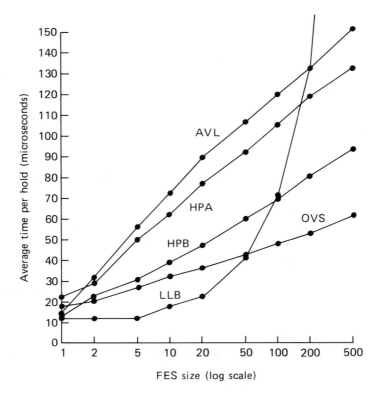

Fig. 3.13 — Hold time for balanced trees, from Evans (1986b). AVL, the AVL tree; HPA, the implicit heap; HPB, the explicit and modified heap; OVS, the binary search tree with overlapping traversal; LLB, the linear list, scanned backwards.

(Reproduced by courtesy of Elsevier Science Publishers, B.V.)

instance, in simulation we are not really concerned with the general problem of merging two priority queues (Françon *et al.*, 1978). Moreover, theoretical studies do not seem able to predict the poor showing of heaps and balanced trees.

Aside from algorithmic complexity studies, there have been several theoretical analyses concerning the expected distribution of FE notices (Vaucher, 1977; Engelbrecht-Wiggans and Maxwell, 1978). If the scheduling distribution probability density function (pdf) is $f(x)$, with cumulative pdf $F(x)$, then at any event occurrence the distribution of FE notices, assuming a steady state has been reached, is

$$g(x) = \frac{1 - F(x)}{a} \quad 0 \leqslant x \leqslant \infty \ ,$$

where $a$ is the mean $E(f(x))$, and the cumulative steady-state distribution is

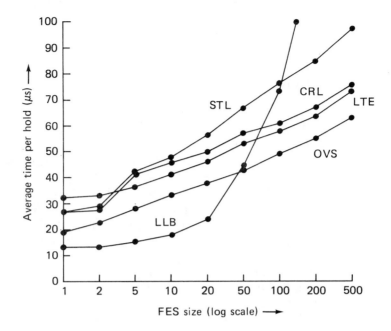

Fig. 3.14 — Hold time for partially balanced trees, from Evans (1986b). Rebalancing
triggers:

STL — invocation-stack length
CRL — count of right and left subtrees
LTE — left subtree empty

(Reproduced by courtesy of Elsevier Science Publishers, B.V.)

$$G(x) = \int_0^x g(t)\, \mathrm{d}t$$

whose mean, $E(G(x))$, called *bias* in Jones (1986) and %*F* in McCormack
(1979) *et seq.*, is a measure of the tendency of the scheduling distribution
towards FIFO or LIFO scheduling. For instance, the constant scheduling
distribution, number (5) in Vaucher and Duval's (1975) list, results in FIFO
scheduling, since each new FE notice is automatically greater than the
maximum in the FES, and its bias is therefore unity. Distribution number
(3), uniform over [0.9,1.1], which is nearly constant, has a bias of 0.967. On
the other hand, the bimodal distribution, which strongly favours samples
nearer the origin, has a bias of 0.13 (Kingston, 1984). The exponential
distribution (1) is unique in that its steady-state distribution of FE notices
$g(x)$ is also exponential, and has a bias of 0.5.

Kingston (1984) performs some analysis on the shapes of the unbalanced
binary search tree and the p-tree. In addition he shows that, for each
distribution, the *entropy* gives the theoretical minimum number of compari-

sons for insertion. Skewed distributions have low entropy, and the lowest entropy corresponds to FIFO scheduling, i.e. distribution number (5). The exponential distribution has maximum entropy.

### 3.5.4 Conclusion

In general, authors of simulation languages tend to be chary about discussing the scheduling algorithm they adopt. According to Engelbrecht-Wiggans and Maxwell (1978), Simscript and later versions of GPSS use linked lists. Nygaard and Dahl (1978) state that Simula uses a p-tree which has $O(1)$ complexity for FIFO and LIFO scheduling, but worst-case insertion could be $O(n)$. GPSS/H (Henriksen and Crain, 1982) naturally uses Henriksen's (1977) algorithm, which is also used in SLAM (Pritsker and Pegden, 1979).

The main problem with the hold model is the lack of decisiveness in the results. Rarely is one scheduling method found to be outstandingly better under all conditions of scheduling distribution. Initialisation of the FES is also problematic, in that in order to be fair one would like each test to start in the same condition, but the steady-state FE-notice distribution for each scheduling distribution is different. Strangely, no one seems to have considered initialising the FES for each scheduling distribution $f(x)$ with the corresponding $g(x)$, the theoretical steady-state distribution of FE notices.

Perhaps the resolution may lie in further consideration of the effective scheduling distribution typical of simulation practice. It is rare that a single scheduling distribution applies to all the events generated in a simulation. It is more likely that each component activity will have its own particular distribution, appropriately parametrised. The future-event set then contains FE notices derived from a *mixture* of distributions. How can this situation be characterised?

An appeal to information theory would suggest that the *exponential distribution* would best characterise the uncertainty of the situation (Mitrani, 1982). It has maximum entropy (Kingston, 1984), and corresponds to a uniform random flux of event occurrences. A simple experiment will verify the tendency of the hold times, obtained from mixtures of scheduling distributions, to those for the exponential (Evans, 1983).

Whichever belief one holds about the expected steady-state distribution of FE notices, whether it will tend towards a FIFO distribution or the exponential, will determine whether the binary search tree is deemed acceptable or not. Assuming it is, the contenders for best scheduling method are Henriksen's algorithm, the binary search tree with overlapping scan and the splay tree. OVS seems especially efficient, as it is so much better (Fig. 3.13) than even the modified heap, which takes short-cuts.

The second paradigm analyses a simulation into three domains. Since scheduling operations are concerned with determining which entity is to move next, they belong to the *allocation* domain, which includes set selection. How is time spent in a simulation? What proportion of a stimulation's total run-time is devoted to scheduling? No one seems to know. Certainly, the larger the FES, the greater proportion of time is spent in scheduling, which means we should prefer a method which handles large FES sizes efficiently. The run-time requirements of each domain are clearly model-dependent.

## 3.6   BRIEF NOTES ON ALGOL 68

### 3.6.1   Aims

In order to make this chapter more self-contained, it was felt that some description of the language Algol 68, in which all the algorithms have been expressed, would be in order. Algol 68 is a powerful language, precisely defined in the Revised Report (van Wijngaarden *et al.*, 1976), so that any description other than this must be treated as superficial and incomplete. Further informal introductions to the language are available in Lindsey and van der Meulen (1977) and McGettrick (1978).

In the introduction, the editors of the Report lay down four main aims of the languages: clarity with completeness, orthogonal design, security and efficiency. These aims seek to provide the programmer with a language conforming to the variously expressed norms of structured programming, in particular the ability to provide facilities with great expressive power based upon a small number of independent primitive concepts. One particular feature of Algol 68, which was found especially useful in specifying operations on data structures, was the **ref** mode concept, which, while sometimes difficult to get right, gives a clear idea of pointer manipulations. The language is therefore well suited to express the details of scheduling algorithms. Some interesting comparisons between Algol 68 and Pascal, probably the most widely known algorithmic language, are given by Berry (1981) and Edwards (1977).

Perhaps the most important point to make about the language from the outset is its adoption of a strict typing policy. A 'static mode checking' ensures that no run-time type (i.e. mode) checking is necessary, except where explicitly demanded. So, every identifier in a program has a definite mode and assignments may be made on the basis of the application of strict rules. This does not imply, however, that an identifier's mode is fixed.

### 3.6.2   Declarations

An important distinction to be made is that between a declaration of an identifier of mode **X** and one of mode **ref X**. For instance the declaration of the integer identifier $i$:

   **int** $i$;

enables $i$ to represent an integer, such as 27, by means of the assignment

   $i := 27$;

but the mode of $i$ is **ref int**, a reference to an item of mode **int**, compared with the constant '27' which has mode **int**. Identifier $i$ is thus capable of referring to an integer, and currently happens to refer to 27. Actually the declaration of *i* is a short-hand for the full form:

**ref int** *i*=**loc int**;

which specifies that identifier *i* has mode **ref int** of local scope. By this declaration, three primitive actions are performed:

- sufficient memory space for the item is reserved;
- the identifier *i* is generated with known mode and scope;
- the identifier is associated with a location.

Consequently, the left-hand side of an assignment, that is the item to be assigned a value, must have a mode which starts with **ref**, otherwise it cannot refer to anything and thus cannot receive an assignment. The right-hand side is then automatically coerced into a mode of the same type, but without the initial **ref**.

Mode **int** is one of several primitive modes of Algol 68. Among others there are **real**, **char**, **bool**, with obvious meanings. Primitive modes may be extended in several ways. Arrays may be composed, enabling a set of values to be indexed, and allowing the array to be treated as a whole. For instance,

[1:3] **int** *triple*;

defines an identifier *triple*, with mode **ref** [ ] **int**, which is an array of three integer references:

*triple*[1], *triple*[2], *triple*[3]

and we can assign

*triple*:=*triad*;

as long as *triad* references, or can be coerced to reference, an array of three integers. Multi-dimensional arrays may be declared analogously. Whenever an array of any mode is declared, the operators **upb** and **lwb** provide the programmer with a means to find the values of upper and lower index bounds. Unlike Pascal, the actual bounds do not form part of the mode.

Arrays are one form of a multiple mode in which each individual element has an identical mode. We are allowed to relax this restriction in the declaration of a record data structure. Suppose we wish to associate a point in the plane in terms of its *x*- and *y*- coordinates with a letter by which it can be identified. It would be convenient to have a triple structure with two **reals** and a **char**. The statement

**struct** (**real** *x,y*, **char** *letter*) *point*;

declares such an identifier, whose position can be accessed by

*x* **of** *point*, *y* **of** *point*

and its identification by

*letter* **of** *point*.

Here *x*, *y*, and *letter* are *field selectors*. When many declarations of the above mode are required, it is more convenient for the programmer to define a new mode, as follows:

**mode place**=**struct** (**real** *x*,*y*, **char** *letter*);

after which a declaration of

**place** *point*;

would have the same effect as the original declaration of *point*. As far as coercions are concerned, the programmer-defined mode **place** is regarded as an 'alternative spelling' of the **struct** definition, and can appear anywhere that a language-defined mode may appear.

We have seen in this chapter how operations on the FES require us to continually delete FE notices and create new ones. If we were to devote an identifier to each, then we would need to invent a method of producing an unlimited number of identifiers, a considerable challenge. This would clearly be very awkward, so we dispense with identifiers and deal instead with *references* to the items, which reside in a special compartment of memory called, rather confusingly, the *heap*. Notice that there is no connection between the heap memory compartment and the algorithm of the same name described in section 3.3.

By not having names, items on the heap may be created and destroyed by explicit instruction, rather than being tied to the declaration block of an identifier and the run-time stack. Referring back to the LLF algorithm of Fig. 3.3, the *insert* procedure, on finding the correct insertion position, creates a new FE notice on the heap, initialises its fields, and assigns references to the new item to the proper fields of the two neighbouring list elements:

#insertion position found#
*pred* **of** *here*:=*succ* **of** *pred* **of** *here*:=
**heap fenotice**:=(*pred* **of** *here*, *here*, *t*)

To destroy a heap item, we need only to remove all references to it, rendering it inaccessible, and to all intents and purposes non-existent. Unlike Pascal, there is no *dispose* function. The memory space occupied by discarded heap items can be collected up and made available for re-use by the garbage collector on one of its occasional reorganisations of the heap.

For performing operations on dynamic data where references to structures are explicitly handled, it is frequently necessary to compare identifiers at the *reference* level, i.e. to test whether two references are referring to the same item. This may be performed using the **is** operator. For instance, consider the following declarations:

**place** *p*;
**ref place** *rp*, *rq*;
*rq*:=*p*;
*rp*:=*rq*;

The first assignment transfers a reference-to-*p* to *rq*, and the second assignment dereferences *rq* once, delivering once more the reference-to-*p* to *rp*. The following test would then discover the fact that *rp* and *p* are identical references:

> **if** *rp* **is** *p* **then** . . .

although

> *rp* **is** *rq*

would yield a **false** value, as these would be compared at the **ref ref** level. In order to explicitly remove one level of **ref**, we may use a *cast* to specify the intended mode of comparison. Thus the test

> **ref place** (*rp*) **is** *rq*

would compare at the single **ref** level, and yield **true**. The **is** operator coerces each of its operands to make their **ref** levels equal.


### 3.6.3   Control structures

The designers of Algol 68 have taken pains to clear up some of the ambiguities of previous versions of Algol by making the control structures conform to a pattern. Particularly, the demarcation between the areas under the influence of different structures is emphasised by means of the 'brackets' **if–fi**, **do–od**, and **case–esac**. In Algol 68, every statement has a value, the value being the result of the final operation, which persists until a semi-colon is met, at which point the value is discarded.

The decision control structure has the familiar form:

> **if** *bool* **then** *section T* **else** *section F* **fi**

where *bool* may be anything, constant, variable, expression, procedure call or string of statements, capable of delivering a **bool** result. When this value is **true**, the set of statements *section T* is executed; if **false** then *section F*. Either section may be empty, in which case the null statement **skip** can be used. Naturally, **else skip** may be omitted.

The iteration control structure is familiar from many programming languages. In Algol 68 there are two main variants: the indexed loop

> **for** *index* **from** *start* **by** *increment* **to** *finish*
> **do**
> > *repeated clause*
> 
> **od**

which defines a fixed number of iterations, and the conditionally terminated loop

> **while** *bool*
> **do**
> > *repeated clause*
> 
> **od**

where the iteration is repeated unless *bool*, which is re-evaluated before starting every new iteration, takes on the value **false**.

If no index is required to service the loop, '**for** *index*' may be omitted. In fact any parts of the **for** clause may be omitted, and the defaults are taken from

> **from** 1 **by** 1 **to** *infinity*

If both **for** and **while** parts are included, the loop may halt for either reason, whichever applies earliest. If no heading is given to a **do–od** bracketed clause, we have an infinite loop

> **do**
> > *repeated clause*
> **od**

which would require the execution of a **goto** in the repeated clause to terminate the program.

The decision control structure based on a **bool** value may be generalised to an *n*-way decision, based upon an integer value. The statement

> **case** *int*
> **in**
> > *section* 1,
> > *section* 2,
> > . . .
> > *section n*
> **out**
> > *section* 0
> **esac**;

evaluates the clause *int* and with its value *i* chooses the *i*th section following **in**. If there is no *i*th section, *section* 0 is chosen. Similarly to the **if** statement, **out skip** may be omitted. A further form of the **case** statement exists to handle identifiers which have modes composed of a **union** of modes. When it is required to find out exactly what is the current mode of an identifier of such an identifier, the **case** statement can define one branch for each mode possibility.

### 3.6.4  Subprograms

The ability to invoke a module of statements at a point in a program is widely regarded as a necessary component of robust and comprehensible programming. In Algol 68 we may define the procedure

> **proc** *aproc*=(*declaration of formal parameters*) *result mode*:
> > *procedure text*;

which can be called by simply mentioning the procedure name at the appropriate position in the program, along with a set of actual parameters

> *aproc* (*actual parameters*);

The actual parameters must correspond one-to-one with the formals as to their mode and order. Declarations of formal parameters are different from ordinary identifier declarations in that there is no abbreviated form allowed. The interpretation of the actual parameters is that they are passed 'by value', enabling values to be passed from the region of the invocation, through the actual parameter list, to the interior of the procedure declaration. In Algol 68, variables are also values, so to be able to assign values to variables in the actual parameter list, they should be declared as a **ref** to the appropriate mode.

In common with all statements, the Algol 68 procedure call has a value, this being formed by the final statement executed in the procedure. If the procedure is being used in a 'function' sense, then this value is the result of the function operation. The mode of the result is given in the procedure declaration as the result mode. When the result is of no interest, as when the procedure is being used as a 'subroutine', the result can be explicitly discarded by making the result mode **void**.

This completes our brief look at Algol 68. We have concentrated on the features which we have used in this chapter, omitting transput, operators, united modes and other advanced features. The language represents a watershed in the development of programming languages, and in a sense has never been bettered. But since its inception, the interests of programmers have changed. By continually exposing languages to face new difficulties, programmers can show up inadequacies of design in practical application. One of the most testing areas in the application of Algol 68 to a practical problem has been the provision of a simulation executive package. For the specification of algorithms the language is fine, but the means of user interaction are found to be rather limiting. We shall discuss these problems more fully in later chapters.

# 4

# Strategies of discrete-event simulation

## 4.1 THE RESOURCE-LIMITED MODEL

As we have seen, the types of systems commonly encountered in management, and studied by Operational Researchers, are quite different from those of physics, chemistry and astronomy. This is reflected in the choice of modelling methods adopted by practitioners in their respective fields. There is a further difference in the inclination of these disciplines towards the models they build. Typically a natural scientist is presented with data from experiments or observations, and seeks a model which will support them. Often the models will take the form of sets of differential or algebraic equations. In management, many systems involve different kinds of phenomena, such as queueing, uncertainty and decision-making, which are not always amenable to a strictly mathematical approach. Also, the component entities which make up the model tend to be more individual, with their own properties. But most importantly, the Operational Researcher is interested in the *design* of a system to achieve a certain performance, rather than merely observing it. This means that the conventional methods of natural science are not readily applicable, so we must resort to simulation.

Instead of trying to develop a theory of how or why a system may exhibit particular phenomena, the simulationist investigates a system in order to experiment with it, or to test out proposed amendments, or even to discover how improvements could be made. What is important here is not so much the investigation of the system as it is, but of what might occur if the system were changed.

The typical management system to be investigated is a generalisation of the queueing model, and involves two kinds of component. Basically, *resources* provide service to a flow of *entities*. From a management point of view, the flowing entities may be customers, whose throughput comprises a source of earning, or perhaps a flow of jobs comprising production in a manufacturing plant. The resources represent a capital investment, so they must be kept busy to realise their full potential

Many problems of management may be seen as special cases of matching the availability of resources to the flow of entities. Assembly lines and job shops are set up to operate efficiently by maximising their throughput. Similarly supermarket check-out tills are manned in such a way as to facilitate customer flow. But we must not forget the costs of increased throughput. In other words, while we would like to prevent the build-up of large queues by installing more check-out tills, we would not want to have check-out operators hanging around idle.

Generally, the *resource-limited model* expresses some kind of trade-off between the supply of resource (which is expensive) and the demand of the entities (whose flow generates income). Besides internal system considerations, there may be a need to try to match external influences, such as the vagaries of demand for production, or customer response to advertising, etc. Not only questions of balance between supply and demand, but questions of policy can be investigated. Similarly, procedures to be undertaken in the case of machine breakdown, maintenance, or emergency may require evaluation before being put into practice.

The OR practitioner observes the system and considers possible improvements which might be put into practice by *simulating* the amended system. Although the resource-limited model is frequently found to be applicable, it by no means exhausts the possibilities of designed systems, and the simulation programming language should be capable of portraying all the various kinds of bottlenecks which may arise. By making observations of the simulation and comparing values of various system properties, such as throughputs, queue build-up, time in system, machine use and so on, the suggested changes may be assessed rationally in a realistic and comprehensive a way as possible.

However, simulations are often quite intricate programs and rarely can one go straight from a conception of a system to its programmatic representation. In the design of a simulation language, we need to adopt a particular simulation *strategy* which affects the methodology and implementation of the simulation program. In this chapter we describe three strategies: event-scheduling, activity-scanning and process-interaction.

## 4.2 EVENT-SCHEDULING

In the two previous chapters we have investigated the details of various time-advance mechanisms which supply the simulation program with a time-ordered sequence of events. Clearly any of these mechanisms could be used as a basis of a rudimentary simulation system, by simply associating with each future-event notice a reference to the program module which describes the actions to be taken whenever an event of this particular type falls due.

### 4.2.1 Methodology

If this approach to the construction of a simulation is adopted, how does the modeller proceed? Essentially the methodology is based upon the identification of the component event modules. The interpretation of the system as resource-limited, by identifying entities and resources, is the first step. Then a list of events which can occur is written down. Each event is taken individually and described in terms of the particular interaction between entity and resource. This description takes the form of a procedure module to be invoked when such an event happens, i.e. when an event notice corresponding to this event-type is retrieved from the future-event set (FES). The event-scheduling approach (or event approach) to simulation is probably the most natural way to proceed, and is frequently adopted when starting from scratch using a general-purpose programming language, like Fortran.

### EXAMPLE: Supermarket

Customers arrive at a supermarket, select their purchases, and then queue for check-out service. The counters are staffed by check-out operators who take each customer in turn, calculate the total cost of the purchase, take the money and wrap the purchased goods, enabling the customer to leave.

Suppose the supervisor is unsure whether employing an extra check-out operator would be beneficial in improving the customer flow. Using simulation, the supervisor may compare the effect of an extra operator with the current situation under identical conditions and make a rational choice, taking costs into consideration. Because the model is executed on the computer, the supervisor is not running the risk of wasting money on unnecessary staff.

How is the supermarket system described from the event-scheduling viewpoint? Firstly, the flowing entities are obviously the *customers*, who, during their transitory existence in the model, go through certain states. These are the selection of purchases, queueing for check-out and participation in service. The resources are composed of the more permanently stationed *check-out operators* who cyclically take on one of two states: serving and idle.

One could argue that the contents of the shelves are also resources. After all, the goods on sale are the reason why customers enter the supermarket! But in this model it is enough to assume that the shelves are always sufficiently well stocked to satisfy customers' expectations, since this is not really relevant to the purpose of the model. The art of deciding just what is relevant to the simulation model depends on many factors: what can be gleaned from observation of the real situation, considerations of privacy, established practice, how much time is available for investigation and so on. Above all, the *purpose* of the model should be borne in mind in deciding questions of detail to be included.

A list of typical events for the supermarket simulation might be:

customer_arrival, start_goods_selection, end_goods_selection, join_check-out_queue, leave_check-out_queue, start_check-out, end_check-out_ and leave_supermarket.

Exactly what constitutes an *event* in the real system is also somewhat arbitrary. One overriding consideration is that the event itself should have zero duration, i.e. it should be instantaneous. So we have the two events 'start_goods_selection' and 'end_goods_selection' rather than just 'goods-_selection' which could not justifiably be thought of as happening instantaneously. On the other hand, we have the event 'customer_arrival' for which it is assumed that the time taken to go through the door is negligible. But in other models, this assumption might not be so tenable.

Going through the door in a lift system, for instance, would not be an ignorable duration because the time taken to load boarding passengers into the lift is obviously going to affect overall journey times, both for passengers boarding and for those continuing their journey. For every passenger, the total journey time will be made up of travel times between floors plus many 'going_through_door' durations exercised by other passengers. So while entering or leaving a supermarket may consume negligible time, a similar action in the case of a lift system may not.

While it may seem at first sight strange to concentrate solely on events, which have zero duration, it should be borne in mind that events represent the occasions when the states of the model change. In fact a running simulation, leaping from one event to the next, busies itself with state-changes at zero-duration happenings, and ignores the periods of resource application between, so that there is a pervasive kind of simulation 'irony' (Spriet and van Steenkiste, 1982) in which we concentrate on describing the instants to the relative detriment of the enduring activities.

Often we find that, in any list of events we may draw up from considering a real system, there are some which turn out to be merely different labels for the same instant. In our example it seems that 'customer_arrival' and 'start_goods_selection' would most likely be coincident, unless we were going to the extent of interposing a 'search_for_basket' activity between them. Similarly 'end_goods_selection' and 'join_check-out_queue' would be coincident, since both indicate the change in customer state from 'goods_selection' to 'queueing'. As an alternative method, rather than identifying the events, we may look at the *states* of the entity (i.e. its *activities*) and derive the instants when they change; it amounts to the same thing.

Having decided on the events, we may turn our attention to the amount of time taken by each activity, or duration for which an entity remains in a state. Taking a commencing event, like 'start_goods_selection', if we can estimate the time for that customer's 'goods_selection' activity, then we may predict the occurrence time of the concluding event 'end_goods_selection' (= 'join_check-out_queue'). This prediction will take the form of a future-event notice and be entered in the future-event set, to await eventual retrieval in due course. Thus a future event is *scheduled* for future activation.

In a simulation, activity time is commonly estimated either by random sampling from a probability distribution, or simply by using a constant value. Random sampling represents uncertainty about a duration; the type of distribution and its parameters must be determined by real-life observation of the system.

### 4.2.2 Reactivations and arrivals

There are some circumstances where scheduling future events is not appropriate. Looking at the event 'join_check-out_queue', we find that if we predicted the event 'start_check-out' (= 'leave_check-out_queue') by random sampling from a probability distribution to estimate queueing time, then the orderly behaviour of the queue would most likely be disrupted. In fact, a customer's duration in a queue is the length of time he has to wait behind others, or, more precisely, the sum of the check-out times for all those who precede him. Practically it is more convenient to model the queue as a set of customers, and simply let the server remove each customer 'first-in, first-out' at the proper time. Thus there is no need to collect data on duration-in-queue times since this information is effectively given by service times.

A similar argument applies to any activity representing a period of idleness. The problem is that an idle resource must be *reactivated* or awoken from its state of idleness by the actions of another entity. In the supermarket we have two states which represent idleness: the queueing customer, and the check-out operators faced with an empty queue. We have already seen how customers may be removed from the queue by the server, reactivating the customer into the state 'being served'. But after a period of slack activity, when the check-out operator is idle and the queue is empty, any new arrival in the queue is deprived of its reactivation mechanism. As a consequence, every customer joining the queue will be condemned to eternal idleness. We can prevent this situation by arranging that the check-out operator is reactivated by the first customer to join an empty queue.

An area where future events are scheduled in a rather special way is that of *generating arrivals*. Normally when we think of duration in an activity or queueing state, we do so in terms of a single entity. Except for successive arrivals into the model, no general relationship will hold between events of different parallel entity flows. To represent an arrival situation, we may observe the occurrence of arrivals in the real system and determine an inter-arrival distribution, or alternatively count the number of arrivals over a fixed period. Then successive arrivals can be modelled by randomly sampling a time interval from the inter-arrival distribution (usually taken to be an exponential distribution) with an appropriate mean value, and scheduling the event 'customer_arrival' for the next customer at this time in the future.

### 4.2.3   Event-scheduling executive
Having decided on the component events, how can we describe in detail the actions to be taken when an event occurs? Let us assume that we are trying to use a conventional programming language for simulation. What additional facilities do we require? A minimal set of requirements that we would need from a simulation package is:

- simulation executive program;
- some means of describing the actions of an event;
- random sampling;
- FES-handling routines;

The task of the executive is to drive the simulation along in its sequence of state-changes; it provides the essential dynamic of the model. More particularly, the executive retrieves the next event from the FES and arranges the appropriate event module to be activated. Leaving aside technicalities, the skeleton of an event-scheduling simulation executive looks like this:

```
mode evnotice=struct (real time, int event_type) ;
evnotice ev_notice ;

bool finished := false ;

while not finished
do
        ev_notice := next event ;
        time := time of ev_notice ;
        event_index := event_type of ev_notice ;
        case event_index in
            ( # event_description 1 # ),
            ( # event_description 2 # ),
                . . .
            ( # event_description n # )
        esac
od
```

Basically we have a cycle executed once per time-beat. Within the cycle there is the retrieval of the next event from the FES, the updating of the simulation-time variable *time*, followed by the choice of a single event module to be executed. The simulation run can be ended at a particular value of time by scheduling an event for that time, and including an assignment of the value **true** to the Boolean variable *finished* in the event module.

Let us now consider some language features for describing the actions performed on event occurrence.

```
entity→state   —puts an entity into a state ;
entity IN state —delivers true if entity is in the state,
                            false otherwise ;
TAKE entity   —remove an entity from a state ;
SCHEDULE event AFTER time
                    —create an event-notice to activate this event
                      at the given time in the future, and insert it
                      into the FES. If the time field is a distribu-
                      tion, then a random sample is taken.
```

### EXAMPLE: Producer–container–consumer
Consider a system comprising three types of entities: producers, containers

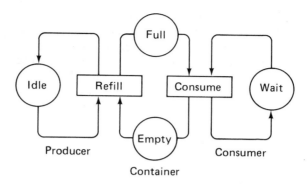

Fig. 4.1 — An activity-cycle diagram for the producer–container–consumer model.

and consumers. Some substance is produced and is delivered to the consumer by containers, which are filled by the producers and emptied by consumers.

This model, outlined in Fig. 4.1, is considerably more abstract than would be formulated in practice, but it has the benefit of highlighting the problem of reactivation. Since all the entity processes are cyclical, we avoid having to schedule arrivals.

Given this abstract, symmetric model, many kinds of resource-limitation are possible, depending on assumptions made about the properties of the various components. For instance, if the number of containers, or their capacity, were too small, then they would cause a resource-limitation. Alternatively, changing the number of producers or consumers, or the time taken in refilling or consumption, could also transform the nature of the resource-limitation. Furthermore, in a simulation, the properties themselves could be transformed at various stages in the evolution of the model, by scheduling the activation of appropriate event modules describing the transformations.

There are also many purposes which could be served by the model. For given numbers of the entities (producers, containers and consumers) we may like to investigate queue build-up at various points, or consider individual producer capabilities, or different customer needs. The scope for expansion to include extra realism is unlimited.

Fig. 4.1 shows the states (activities) entered by the entities as they flow through the system, in an *activity-cycle diagram*. There are two types of states, represented by boxes and circles, which indicate whether or not there is any resource-limitation on their start-up. For example, the 'refill' state requires the availability of both producer and container. If one or other is busy, then the 'refill' activity will be delayed. On the completion of 'refill', the container takes on the state 'full', and the producer becomes 'idle'.

While the boxes represent activities which will take a determinable amount of time, the circles are 'do-nothing' states where entities wait for an indefinite time to be reactivated. We may model such states as queues, or as sets of idle entities, whichever is appropriate. Minimally, we need to model only the number of items in such states.

Starting from the main activities of the model we can see that the pivotal events are given by the beginning and end of each activity:

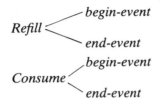

Taking each event in turn, we can write down the specific details of the event description:

> EVENT *refill . begin* :
> > IF *container* IN *empty*
> > AND *producer* IN *idle*
> > THEN
> > > TAKE *container, producer*
> > > *container→refill*
> > > *producer→refill*
> > > SCHEDULE *refill . end* AFTER *refill*
> >
> > FI

The event description is conditional on the presence of two entities in specific states; this condition is thus naturally at the head of the description. If the condition is satisfied, a container and a producer are taken and their states changed to the activity name 'refill'. Finally the end of the activity is scheduled at a future event with an occurrence time obtained from a random sample from a distribution also named 'refill'. In general, for beginning-events we deal with the conditions for start-up and the scheduling of the ending.

The other begin-event is described similarly:

> EVENT *consume . begin* :
> > IF *container* IN *full*
> > AND *customer* IN *wait*
> > THEN
> > > TAKE *container, customer*
> > > *container→consume*
> > > *consumer→consume*
> > > SCHEDULE *consume . end* AFTER *consume*
> >
> > FI

The end-events are simpler, or appear so at first, since they represent entry into states in which the entities are merely passive. There is no conditional start-up, and entities leave these states by reactivation from elsewhere. It seems that it is sufficient simply to write

> EVENT *refill . end*
>    ( *producer→idle*
>      *container→full* )

and

> EVENT *consume . end*
>    ( *container→empty*
>      *consumer→wait* )

but a little reflection will reveal that the system as a whole is incomplete because no begin-events are scheduled. Thus the model contains no means of re-starting either activity, even assuming that they are started at all. The event modules are procedures called by event-notices retrieved from the FES. If there is no event-notice, then there is no activation of the corresponding event module. The solution is to reactivate the begin-events immediately any entities in their start-up conditions become available. Because of the *conditional* nature of the begin-events, reactivation simply means re-try the condition: the body of the description is entered only if the condition permits. Thus the end-events become:

> EVENT *refill . end*
>    ( *producer→idle*
>      *container→full*
>      SCHEDULE *refill . begin* AT *now*
>      SCHEDULE *consume . begin* AT *now* )

and

> EVENT *consume . end*
>    ( *container→empty*
>      *consumer→wait*
>      SCHEDULE *refill . begin* AT *now*
>      SCHEDULE *consume . begin* AT *now* )

Scheduling AT now causes an event to be subsequently reactivated, but at the current value of simulated time. It can be regarded as a 'nudging signal' to broadcast information to conditional events that certain other events have happened. Here, two reactivations must be called in each end-event, because in each case the entities entering their passive states are mentioned in the conditionals of two begin-events. Instead of one time-beat for each new value of *time*, we now have three. We can see that the problem of reactivation considerably complicates the structure of the model, and the programmer should be keenly aware of the interdependency between event descriptions.

From the viewpoint of program structure, the event-scheduling approach to simulation leads to a modularisation of the program in terms of interdependent parts. Thus changing any part of the program, involving say a single event, could require adjustments in many other event modules which refer to, or are referred to by, the event description in question. In general, making any change to one event could conceivably necessitate adjustments in all other modules. Thus the model will be difficult to amend. This is a considerable drawback, since one of the primary reasons for simulation at all is to make it possible to experiment with the model under different conditions of operation. It is primarily for this reason that the event-scheduling approach has lost popularity.

## 4.3   ACTIVITY-SCANNING

We have seen that the need to reactivate events by incorporating nudging signals somewhat spoils the simple approach of straightforward event-scheduling. A more effective mechanism for handling reactivations, especially when more complex models are envisaged, would greatly improve the power of simulation. To meet this problem, the *activity-scanning* strategy was proposed (Tocher and Owen, 1960). The methodological aspects of this approach are similar to those of the event-scheduling approach. As before, a list of events is drawn up, but they are further classified into two sets of events, confusingly called B-activities and C-activities. B-activities are events athat are *Bound* (i.e. certain) to happen and C-activities are those that happen *Conditionally*, giving rise to two mutually exclusive sets of events.

Before continuing, we should first note the ambiguous use of the word 'activity'. Previously we have used it to represent a significant duration of time, begun and ended by discrete events. But in the activity-scanning approach, an activity is the name of an event description, of either B- or C-type. Being an event, a B- or C-activity occurs instantaneously, i.e. it has zero duration, and refers to the actions of the program in describing the change of state. To avoid confusion, we shall distinguish between the two meanings by preceding the work 'activity' by B- or C- to denote an event description.

Applying the activity methodology to the producer–container–consumer model means that our original set of events is now partitioned into two:

> B-activities :
>> *end . refill*
>> *end . consume* ;
> C-activities :
>> *begin . refill*
>> *begin . consume* ;

Generally, a C-activity can be identified by considering the question 'under what conditions will it take place?'. If an entity in a certain state is necessary, i.e. if the event is state-dependent, then the event should be described in a C-activity. The descriptions of the event actions remain the same, except that no nudging signals are necessary: nudging reactivation is now undertaken by the simulation executive. So our end-events resume their simple form:

> BACT *refill . end*
>    ( *producer*→*idle*
>     *container*→*full* )

and

> BACT *consume . end*
>    ( *container*→*empty*
>     *consumer*→*wait* )

Because they usually describe entities merely taking up passive states, B-activities are called 'bookkeeping activities' by some writers, whereas the C-activities comprise begin-events and handle the more significant job of expressing resource-limitation in the model.

    When considering whether or not an event is conditional, it should be remembered that we are dealing with discrete state-changes only. In this example one might be tempted to describe 'refill . end' as conditional on the container becoming full. This is incorrect, however, since the container can take only two states, 'empty' and 'full', and the amount is not allowed to change continuously. In a purely discrete model, the container becoming full is therefore a *consequence* of the 'refill . end' B-activity.

    The control structure of the three-phase activity-scanning simulation executive is

```
mode cact = proc void ;
[1 : m] cact c ;                    # the array of C-activities #
while not finished
do
    ev_notice := next event ;
    time := time of ev_notice;              # the A-phase #
    B_index := event_type of ev_notice;
    case B_index in
        ( # B-activity 1 # ),
        ( # B-activity 2 # ),
            ...                             # the B-phase #
        ( # B-activity n # ),
    esac;
    for i to m do c[i] od               # the C-phase #
od
```

The simulation executive for the activity approach is composed of three phases:

A — the time-Advance phase;
B — the execution of a single B-activity;
C — calling each C-activity, collected in the array $c$.

Each time-beat consists of one cycle through the **do**-loop, executing the A-, B- and C-phases in succession. Alternative names for this strategy are the 'three-phase' or 'ABC' approach. The time-advance phase is the same as that of the event-scheduling strategy, but the *event type* delivered from the retrieved future-event notice is now an index to a B-activity. After execution of this one B-activity, the whole collection of C-activities in the array $c$ is examined to see if their conditions, based on the model state just modified by the B-activity, will allow entry.

The basic structure of each C-activity is

- a preliminary 'test-head';
- the action part, called the 'body'.

The *test-head* contains the condition which must evaluate to **true** before access is allowed into the *body* of the C-activity. While the B-activities just refer to a simple event with no concern for any consequences, each C-activity contains all the conditions for a state-dependent event followed by its actions. Typically the begin-events of the event-scheduling strategy conform to the pattern of the C-activity, although in more complex models it is possible to have conditional end-events.

The effect of an event occurrence, as triggered by a retrieval from the FES, may be regarded as consisting of two parts. Firstly, there are the *immediate effects*, such as changes of state of various entities. These effects are described in B-activities. Secondly, there are *consequences* of these state-changes, typically in the form of entity reactivations. Since each C-activity consists of an event description, preceded by its preconditions, an activation of every C-activity will enable the nudging signal to be broadcast automatically. Since each activity defines its own actions and its own conditions, the resulting set of activities is more modular than was the case for the event descriptions.

For the supermarket simulation, the B- and C-activities are

> B-activities
> *customer arrival*
> *join check-out queue*
> *end check-out*
>
> C-activities:
> *start check-out*

In all, there are three B-activities and only one C-activity for this simple model. The conditions for *start check-out* are

> *check-out operator non-busy, and*
> *number of waiting customers > 0.*

***Exercise***
Using the notation of the previous section, write down the complete set of B- and C-activities for the supermarket model. You will need to invent some notation for creating and destroying entities to represent customer arrival and departure. Creations should occur according to a specified inter-arrival distribution.

In a more complex model. we may well wish to include extra preconditions on the event 'start_check-out', such as considering a condition 'not_end_of_ shift'. Extra conditions could be either inserted alongside existing ones, or incorporated in the test-head of another C-activity; it depends on whether different actions are required. So we can see that having all the conditions together which pertain to one set of actions makes the simulation modules more independent and makes it easier to adapt them to changes in the real system, or to incorporate a greater amount of detail.

However, there are disadvantages with a C-phase. It is rather wasteful to keep checking a rare condition on every time-beat. It may be known that only a particular B-activity could ever set up the conditions for a C-activity body to be entered. But it is difficult to incorporate such information in a general way.

Attempts have been made (de Carvalho and Crookes, 1976; Clementson, 1973) to link together loosely into the same cell, sets of B- and C-activities which share entities. Thus a set of C-activities is grouped with all the B-activities from which they obtain their entities. On each time-beat the set of C-activities to be examined is restricted to those which are grouped with the B-activity executed.

It seems natural to group together B- and C-activities pertaining to a common entity, and to order them in their actual sequence of succession. If we do this, we are a step nearer the concept of a sequence of happenings for a single entity; the third strategy to modelling takes precisely this viewpoint.

## 4.4   PROCESS-INTERACTION

We have seen that both the event-scheduling and the activity-scanning strategies take as their starting-point a set of event modules. The set of events is entirely amorphous, being abstracted from a broad view taken over the whole system. Any sequential relationship which existed between the events must be explicitly included in the event-scheduling methodology, whereas in the activity-scanning approach we discriminate between immediate effects (B-activities) and consequent effects (C-activities).

But when we look at any list of events describing the system, we frequently find that the order in which they are written down reflects some kind of sequence which applies to the order of their occurrence in the *entity flow*. The ordering of events comes about by the modeller imagining himself in the position of a typical entity passing through the system, and is evident in the list of events describing the supermarket (see above). With the two methodologies we have met so far, this readily available sequence information is effectively ignored, whereas the process-interaction approach capitalises upon it.

In any system where there are parallel flows of entities, the meaning of 'order of events' is rather ambiguous. From the viewpoint of a customer in a supermarket, the 'start_check-out' event certainly follows the 'join_check-out_queue' event, probably after a delay of some time. But taking an overall perspective, without attaching to any particular customer, the event activations seem to occur in a quite undetermined, random order. In order to include information about event sequence, we must therefore concentrate our attention upon a single entity.

The process-interaction methodology describes the system's workings from the viewpoint of an entity flowing through the system. The model is thus described as a projected life-history of a typical entity, called a *process*. The process defines the entity's resource requirements, its possible impediments, and its duration in each of its activities. Each new instance of an entity, representing a separate parallel flow, takes a copy of the process as a guide for its development. Although the events will occur in a specific sequence embedded in the process, the running simulation contains many instances of the process existing at different stages of development, giving rise to the overall haphazard pattern of event occurrence.

The execution of a process simulation takes a single process instance and executes it until a point is reached when it partakes in an activity; in other words, until the entity description expresses the passage of time. As usual, at the beginning of an activity, the end-event is scheduled. The event-notice includes a pointer to the position in the instance where it had reached in its progression, identifying the subsequent *reactivation point*. Next, another instance is taken, reactivated and progressed in a similar way. The choice of which entity to reactivate is determined by the retrieved future-event notice, which will correspond to the most imminent event occurrence. On reactivation, an instance is continued from the point where it was last discontinued.

The modularisation achieved in the process-interaction methodology is based on the sequence of events and activities comprising a process. In general, processes may apply to both entities and resources, but quite often the resources are sufficiently trivial (just busy–idle cycles) to be treated simply. In the case where different kinds of entity are involved, or the behaviour of a resource is such that it demands a full process description, then several interacting processes will be needed to define the whole system.

For events and activities, the *subroutine* control structure is adequate to describe their mode of activation, directed by retrievals from the FES, or, in the case of C-activities, from the C-phase of the activity executive. This is because, on being invoked, the whole event or activity is executed. But for processes, we need to be able to discontinue and re-continue within a single module. To accomplish this pattern of invocation, the process's behaviour is mapped onto a control structure known as the *coroutine*.

### 4.4.1 Coroutines and subroutines

In conventional programming, large programs are modularised in terms of subroutines, and their interrelationships are thereby restricted to a top-down invocation hierarchy. Semantically, a subroutine call is equivalent to the insertion of a piece of code, the subroutine text, at the point of invocation. The subroutine text is regarded by the invoking subprogram as a whole. The reason for using a subroutine may be because the same actions are required elsewhere and we wish to avoid duplication, or maybe a subroutine pattern decomposes the program into something easier to comprehend.

The main text invoking a subroutine is rather like a busy bureaucrat finding something in his daily work that is too tedious or mundane for his own attention, delegating responsibility for it to a mere functionary, with the proviso that he should report back when the work is complete. Fig. 4.2 indicates the flow of control in a subroutine invocation.

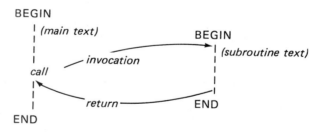

Fig. 4.2 — Subroutine invocation hierarchy.

A subroutine may call another subroutine, or itself, extending the pattern in an unlimited hierarchy. We should note four properties of subroutines;

- they are always entered from the beginning;
- control ultimately returns to the invocation point;
- the return address is the property of the invoked routine;
- the invocation pattern is hierarchical.

Coroutines are similar to subroutines in that they serve as program modules, but they possess a totally different invocation pattern (Fig. 4.3).

Coroutines are invoked by *resume*, which differs from the calling of a subroutine in that the resumed coroutine is started from the place where its execution was last quitted, which is not necessarily its textual beginning. Once a coroutine is discontinued by resuming another, there is no guarantee

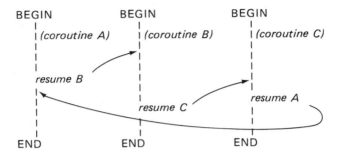

Fig. 4.3 — Pattern of coroutine invocation.

that control will ever return. Compared with subroutines, there are four corresponding properties of coroutines:

- they are entered at the reactivation point;
- the destination of control is determinable by the routine;
- the routine keeps its own reactivation-point record;
- the invocation pattern is heterarchical.

Sets of coroutines do not therefore possess any master–slave interrelationship, as do subroutines. Commonplace analogies are more difficult to construct, but if we had to check the accuracy of translation of a book in language $X$ to language $Y$, we might proceed in a 'coroutine resumption' manner. Let the books be partitioned into paragraphs $\{X_i\}$ and $\{Y_i\}$. Suppose we read paragraph $X_i$, then to check its translation, we place a bookmark next to $X_{i+1}$ and turn to the $Y$ book at its marked paragraph, say $Y_i$, which is read and compared with the remembered $X_i$. After comparison, $Y$ is marked at $Y_{i+1}$ and $X$ is resumed. The switching between books, and the updating of the pointers, is directly analogous to two coroutines with mutual resumption.

Using a loose sociological analogy, we can say that the subroutine hierarchy corresponds to an autocracy, because there must be a unique routine (the master program) to which control will eventually revert, whereas a system of coroutines is like a democracy where each individual 'does his own thing', and none has ultimate control.

It is interesting to observe that while subroutines are very widely used in programming, few simple examples of coroutines actually exist, even though the structure is clearly very powerful. Knuth (1973a) also notes this curious fact and Bornat (1979) thinks their real power is yet to be revealed. Since then, a few more applications of coroutines have come to light (Allison, 1983).

### 4.4.2   Coroutines in simulation

The process view of simulation makes use of the coroutine idea in a slightly restrained form called the *semi-coroutine* (Dahl, 1968), or *generator*, as Marlin (1980) prefers, in analogy to random-number generators. Most of the properties of the coroutine are retained, but the ability to choose which other coroutine is to be resumed is largely vested in a superior module, which makes the selection according to the next-event notice taken from the FES. Specific resumption of process instances corresponds to activations which arise from state-dependent events.

The model-building technique for process-interaction starts, as usual, from an identification of resources and flowing entities. Then the modeller imagines himself as the entity and describes the normal procession through stages of interaction with the resources. To describe the interaction, the language might contain the following set of interaction primitives:

GEN *entity* EVERY *inter_arrival*
— the generation (arrival) of an *entity* according to a given inter-arrival distribution;

NEG *entity*  — termination (departure) of the *entity*;

ACQ *resource*  — acquisition of a *resource*. If the resource is not available when the acquisition is attempted, then the entity will wait in a FIFO queue until the acquisition is successful;

REL *resource*  — relinquishing an acquired *resource*, which can be acquired by another entity;

TIME *duration*  — denotes the persistence of an activity (state) for a *duration*.

The supermarket simulation expressed in process-interaction terms is simply

```
GEN    customer EVERY arr
TIME   goods_selection
ACQ    check-out_operator
TIME   check-out
REL    check-out_operator
NEG    customer
```

What is immediately striking about this version of our model is its conciseness. Since there is only one entity-type — the customer — the whole model reduces to a single process. The resource is mentioned only as a parameter to the ACQ and REL operations. If there were more than a single resource then the number would have to be declared in advance.

In GEN the inter-arrival distribution *arr* is specified, to control the successive arrivals of each customer. On arrival, an instance of the customer process is created and executed, statement by statement, until an activity is encountered. An activity simply models the passage of time, either a period of working or a period of idleness. The

> TIME *goods_selection*

statement, causes an 'end_goods_selection' event to be inserted into the FES. The event will occur at a point in future time determined by a sample from the distribution *goods_selection*. Meanwhile the original process instance becomes quiescent in a deactivated state, and the system will retrieve the next event and reactivate another process instance. When the end 'goods_selection' event is later retrieved, the process instance will be reactivated from the ACQ statement.

Acquisitions, when successful, happen instantaneously. But if there is no resource available then the process becomes once more deactivated and placed in a queue, to be subsequently reactivated by the check-out operator, servicing the queue in FIFO order. If the queue empties then the check-out operator becomes deactivated, to be reactivated later by the arrival of the next customer. All this activation and deactivation can be carried out behind the scenes, by the process-interaction simulation executive. Once acquired, the check-out operator and the customer are consigned to a further deactivation, modelling the check-out activity. When this period has elapsed, the check-out operator is released, and made available to other acquisitions.

The NEG statement causes a deletion of the customer from the model, which entails the destruction of its process instance from the program. The elapse of model time between arrival (GEN) and departure (NEG) is called the *transit time* and is a useful indication of efficiency of the supermarket, as seen from the customer's angle. Other measures of efficiency are the average queue length, server utilisation, proportion of queue entrants passing through in zero time, etc. All such data collection can be performed automatically by the process-interaction simulation executive.

By taking the individual entry viewpoint, and representing the model as a set of interacting processes, enables the executive program to subsume responsibility for handling many reactivations, and thereby to obtain a higher-level program as a result.

## 4.5   COMPARISON OF STRATEGIES

### 4.5.1   Summary

Few topics are more contentious when simulationists meet than the subject of simulation strategy, variously called 'approach', 'world view' (Zeigler, 1976), or 'decomposition methodology' (Spriet and Vansteenkiste, 1982), *etc.* Whatever strategy is chosen by a simulation language in its design phase, it remains a permanent characteristic of the language, because the strategy chosen determines the algorithm for the simulation executive program, thereby affecting the way in which the model must be presented to the computer. From the user's point of view, the world view of the simulation language underlies the interface between modeller and computer, and affects, either by hindering or facilitating, how the modeller views the world. So the superiority of each strategy is keenly disputed.

It is not clear whether there are any differences between strategies as regards their 'universality' (Zeigler, 1976), i.e. whether there are any models which are not representable by a particular strategy. Since little work has been done on this question, strategies are generally assumed to be equally powerful, and their differences have become matters of convenience, security, efficiency and familiarity.

Many comparisons are marred by the inclusion of language features or implementation details. Few implementations of strategies are 'pure' and language advocates will blur distinctions wherever there seems an advantage in doing so. Languages are concerned also with features which are relatively independent of strategy, such as output facilities and data collection. Language details will be more thoroughly assessed in Chapter 6.

Nevertheless, despite the barely concealed commercialism of some language comparisons, a few valuable comparisons of strategies have been made (Hills, 1973; Zeigler, 1976; Hooper and Reilly, 1982; Birtwistle *et al.* 1985). Hills (1973) programs the same model using each of the three strategies, written in Simula. He compares length of program, run-time, and ease of programming. Zeigler (1976) takes a more descriptive, analytical view, which is taken further by Hooper and Reilly (1982), who end up with a comprehensive list with eleven points of comparison. Unfortunately they seem to concentrate too closely on GPSS as an exemplar of the process-interaction strategy.

Most comparisons come to fairly equivocal conclusions, with a slight preference for process-interaction. But proponents of activity-scanning (e.g. Davies, 1979) maintain that their strategy is far superior for the more practical, interrogative style of specifying models. It seems unlikely whether there could ever be a fair comparison such that one strategy could be pronounced 'the best', in all respects. To do this the *typical* simulation would have to be advanced, together with the *typical* programming environment, the phantom-like existences of which are doubtful.

### 4.5.2   Event vs. activity

We may summarise diagramatically the main aspects of these two strategies in Fig. 4.4. Suppose that an entity state is to change from *state X* to *state Y* at a time $t$. A future-event notice pertaining to this state-change is put in the FES. When the future-event notice is retrieved, the current time is set to $t$ and the actioning of the state-change is carried out. The method of performing the state-change is the point in which the two strategies differ most.

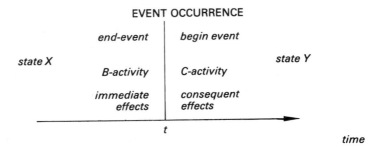

Fig. 4.4 — Relationship of event and activity on occurrence.

In the event-scheduling strategy the $X.end$ EVENT will place the entity in a passive state, then SCHEDULE, possibly alongside other begin-events, the $Y.begin$ EVENT to occur at the current value of time, $t$. In the activity-scanning strategy the invocation of the $Y.begin$ CACT module is performed automatically on every time-beat, by the C-phase. While the activity-scanning strategy separates neatly the immediate and consequential phenomena of an event, the event-scheduling strategy tries to associate both kinds in a single module.

The event strategy has already been criticised for leaving the responsibility of prescheduling all necessary reactivations to the programmer. Using an event-strategy language, it is all too easy to write a simulation which 'works' in the sense that it produces apparently meaningful output, but fails to capture accurately the true workings of the reality in one or more critical areas. The activity-scanning strategy attempts to solve this by invoking all conditional event modules after every state-change caused by a B-activity. While the activity strategy is a safeguard, it tends to be rather inefficient. An event-scheduling simulation with a complete set of reactivations would be more efficient at run-time, but would be more difficult to program. To some extent the simulationist pays for security.

The difficulty with opting for efficiency is that it is all too easy to overlook the inclusion of the nudging signals so that bottlenecks may arise in the model which have no counterpart in the real situation. Furthermore, the interdependency of the event modules makes the simulation difficult to understand. It is primarily for these reasons that the event strategy is no longer popular.

Nor is the activity-scanning strategy foolproof. Many reactivation opportunities can be detected after the state-changes of the B-phase by scanning the C-activities, but what of the state-changes brought about by the C-phase? Any C-activity entered may perform a state-change such that a previously occupied resource becomes idle, enabling an activity to start up. But the availability of the resource will not be detected until the next C-phase, which will most likely happen at a later value of simulated time. For simple models, where activity start-up is modelled by C-activity and activity completion by B-activity, there is no problem, but for activities which end conditionally, the modeller must remain alert to this difficulty.

One answer is to repeat the C-phase or, more completely, to repeat the C-phase until no more reactivations are possible, or until no C-activity test-head allows entry. Unfortunately it is very difficult to guarantee that this repeat loop will ever terminate. One can imagine a situation where, for a particular state of the model, one C-activity releases an entity and another claims it, and in doing so sets up the condition for the release of the same entity once more. Such a modelling error would be difficult to detect, and for this reason no one seems to recommend C-phase repetition. Once more, it is left to the modeller to take precautions, such as performing the release of the entity in a B-activity scheduled AT now.

Another problem with the C-phase which is rarely touched upon is the *order* of scanning the C-activities. The C-activities are presented to the executive in some order and this usually becomes the order in which their test-heads are scanned. Most likely the modeller is unaware that he is implying any sequential relationship between the C-activities when he assigns them to the $c$ array. Suppose we have two C-activities $c_1$ and $c_2$ which have the same test-heads. In every C-phase, $c_1$ will get the first consideration for deployment of the resources, and $c_2$ will be satisfied only if $c_1$ is. In some cases this type of distribution occurs in reality, but usually it arises as an artefact of the C-phase.

This problem of 'fair' distribution is most explicit in the activity-scanning strategy. It is emphasised because the executive is continually trying out all conditional events. Sequencing of events is explicit for the event-scheduling strategy and implicit in the process-interaction strategy, so the activity-scanning strategy is more flexible, and can more easily base decisions on a system-wide perspective, especially since the C-activity test-heads may contain Boolean expressions in terms of *any* system properties. Because of this, the fairness problem is readily apparent. For other strategies these considerations do not readily arise, but one suspects that they are as guilty of unfairness, although in a more obscure form.

One solution would be to interpose an extra phase between the testing and actioning of the C-activities. Three scans would then replace the conventional C-phase. In the first scan, C-activities would be selected if their test-heads evaluate to **true**. Secondly, the selected C-activities are examined to decide on the allocation of available resources. Finally the bodies of the C-activities to which resources are to be made available are executed. However the daunting prospect of writing such an intricate and obscure piece of program to describe an equitable distribution in all possible circumstances would deter the majority of simulationists, so the problems of unfair allocation are swept under the carpet. These points are considered in more detail in Chapter 8, where the engagement strategy is introduced. Interestingly, a similar three-part schema for an abstract characterisation of computation organisation has been suggested (Treleaven *et al.*, 1982).

Occasionally, C-phase priority schemes are put forward (Clementson, 1977) so as to give each C-activity an explicit position in the pecking order. The problem with C-activity priority is that in reality we require the priority to be dependent on the nature of the available resources being distributed, so a once-and-for-all assignment of priority is inappropriate, except for designed systems which really employ fixed priorities, such as computer operating systems.

In situations like a time-sharing computer system, where fairness is an important practical problem, and contentions are rife because of the rigid and centralised synchronisation of the system, assigning priorities to operations is commonplace. However, in more natural systems, such allocations are achieved by human decision-making or the application of a complex set of rules, or by simply leaving the result to chance. Humans seem better able to take an overall view, and to proceed with a course of action for the 'well-being' of the system.

In general, less attention is paid to the activity-scanning strategy than to the other two, in part because of its early two-phase implementation, in which each resource has its own time cell which records the time at which the resource is due to change state. If the time cell is less than the simulation clock, the resource is waiting idle. Time advance is by means of scanning the whole set of resources with time cell greater than or equal to the current clock time, searching for the minimum. This identifies the next value of simulated time.

With the new value of clock time, all activities are scanned, and the ones that allow entry are activated. Activity-scanning is repeated until no more action is possible, when the next time advance by resource-scan is carried out. The original activity-scanning executive consists of two phases:

- resource-scan for minimum positive time value;
- repeated activity-scan.

The modern activity-strategy, the three-phase strategy, is really a hybrid (Zeigler, 1976), using a next-event algorithm from the event-scheduling strategy for time advance, and retaining the C-phase conditional event-scan from the original activity-scanning strategy.

An interesting early comparison of these strategies is given in Laski (1965). Most comparisons seem to concur with Hills (1973) in the benefits of activity-scanning over pure event-scheduling, especially when there is high interdependence between events, or when testing of the global model state is necessary, or if changes in the control of the real system are the object of the study. Event-scheduling, on the other hand, is repetitious in its specification of control and is not easy to comprehend; but, once written it has higher run-time efficiency.

Further consideration of the activity strategy, along with activity-cycle diagrams, will be given in Chapter 6.

### 4.5.3 Hierarchy vs. heterarchy

Perhaps the process-interaction strategy can combine conciseness of description with efficiency of execution? Coroutine resumption is a facility provided by few common languages, so the idea is unfamiliar to many programmers. The invocation of a routine at a point other than its start seems to go against the ideas of structured programming. But if ever the coroutine ceases to be so unfamiliar a control structure, maybe the stategy would lose its strangeness. It is certainly true that the process incorporates more of the available information about the sequential nature of the events, as seen from the viewpoint of the individual entity. However, the viewpoint of the entity itself is not always the best to describe the functioning of the system.

There is an English saying 'The onlooker sees most of the game', while on the other hand many spectators often feel that if only they could be on the field they would get a better appreciation of events. It seems that for the description of any system there are two mutually complementary perspectives: that of the performer and that of the onlooker. Brecht (1957), in his analysis of theatre, came to the same conclusion. In epic theatre, the spectator is an observer of spectacle, whereas drama closely involves the audience in the proceedings on stage. Viewpoints can be either close and involved, or wide but remote.

If we propose to model the system by the sequence of happenings to a typical entity, then we assume that the whole model is encompassed by the entity. Frequently we find that such an assumption is untenable; in complex models the resource often undergoes a significant life-cycle development. Instead of a rigid distinction beteen flowing entity and stationary resource, we need to model both as processes *cooperating* (Birtwistle, 1979) in activities.

The process-interaction strategy solves the activation difficulties of event-scheduling by describing the model as a *sequence of activities*. If resources are modelled in this way we find that a cyclical form of process is often more useful. Where resources may not always proceed in the same sequence of activities, as in this case of a multi-function resource such as a crane, then the crane returns to an idle state on completion of an activity. The occurrence of a need for the crane arising elsewhere can reactivate it into the appropriate activity. It seems that the assumption of sequential activities does not in effect preclude the expression of any model.

But there is one particular difference which separates the process strategy from others: that is the problem of deadlock. Consider the following process (adapted from Birtwistle, 1979):

```
GEN   ship EVERY 80
ACQ   tug
ACQ   berth
TIME  30              # enter harbour #
REL   tug
TIME  60              # unload #
ACQ   tug
REL   berth
TIME  60              # leave harbour #
REL   tug
NEG   ship
```

Superficially this could be taken for a reasonable description of a docking procedure: a ship arrives, acquires tug and berth (of which there is one of each), enters the harbour, releases the tug, unloads, re-acquires the tug, and leaves. Let us analyse it by proposing a scenario:

time=0   ship1 acquires tug and berth and enters harbour;
time=30  ship1 releases tug and commences unloading;
time=80  ship2 acquires tug and queues for berth;
time=90  ship1 finishes unloading and queues for tug;

At this point the system is deadlocked; it can progress no further since each ship has what the other wants and neither can release these resources before they have acquired the other. Ships will continue to be generated but will not get further than joining the queue for berthing. Obviously the model is not mirroring what would happen in reality.

In the present case, the situation may be amended simply by interchanging the second and third lines, the acquiring of tug and berth; and we might blame the modeller for not spotting this. If we now look at the textual pattern of resource acquisition, we find that the ACQ–REL brackets for each resource do not overlap; however, this is not a general guide to absence of potential deadlock, since it would be inapplicable in the case of separate processes. Overall acquisition pattern can be very complicated, and the possibility of deadlock in general is difficult to detect.

On the other hand the annoyed modeller might justifiably complain that the language is unduly miserly in allowing the acquisition of only one resource per ACQ statement. If he could acquire both tug and berth on an all-or-nothing basis, then there would be no deadlock and this would more closely imitate the real acquisition procedure.

Using the event-scheduling and activity-scanning strategies the problem of deadlock does not arise in such a clear-cut form; it stems from the strong emphasis on the sequential nature of the process, and the assumption that each resource-acquisition is a separate action. These properties are most prevalent in computer systems and the process strategy is a popular choice for simulating operating systems, computer networks, etc.

Strangely, the importance of sequence of activities and coroutine activation in the process-interaction strategy seems to have escaped some commentators. Both Zeigler (1976) and Spriet and Vansteenkiste (1982) state that the process-interaction strategy is a combination of event-scheduling and activity-scanning. Once more, the early comparison of Hills (1973) is comprehensive. The main advantages are conciseness of expression and run-time efficiency, but against these, Hills claims that a process-interaction model is hard to specify and modify. Zeigler (1976) is also enthusiastic for process-interaction, and claims a higher likelihood of correct implementation and hence quicker debugging. Another advantage (Birtwistle *et al.*, 1985) is that the process is a more unified construct than a set of subroutines, so that models can be more readily stored in a library.

In Chapter 6 we suggest a mode of comparison based on the information preserved by transforming from a strategy-independent network according to each strategy. Further consideration of strategy will be made in Chapter 8, where the *engagement* strategy is put forward.

# 5

# Simulating chance

## 5.1 NATURE OF CHANCE

While the approach to simulation taken here does not consider random phenomena to be the main focus of interest in discrete-event simulation, it is undeniable that randomness pervades the whole area. One reason is that the simulationist is usually interested in the response of systems to workloads and pressures which are uncoordinated and arbitrary. There is a feeling that, in the modelling of designed systems, such influences are more realistic if modelled with an element of chance. However, the acceptance of randomness into the model means we are faced with an immediate problem in the interpretation of simulation results: are they really significant or due merely to chance? We can attempt to answer this question, and thus be able to simulate more effectively, by taking steps to ensure model validity and adopting an astute experimental design.

Early books on simulation, starting with Tocher (1963), laid great stress on the role of randomness. Given the wartime origins of Operational Research in battlefield simulations, where each protagonist must be prepared for any eventuality, the determinacy which went along with traditional applications of mathematics was no longer thought appropriate. Simulations were at that time regarded simply as models involving a probability term. But, with the gradual trend of computer languages to become more expressive and less machine-oriented, this view has gradually changed in favour of increasing emphasis in capturing the dynamic of the system in the control structures of a simulation program, e.g. Kreutzer (1986).

Let us first look at types of phenomena which are regarded as being 'random'. Firstly, what is meant by the term? There seems to be no completely satisfactory definition. In fact the phrase 'at random' is sometimes used in such a vague way as to be meaningless. Is 2 a random number? There is no way of knowing, unless we know more about the circumstances of its occurrence. Many discussions go astray on this point. Schoffeniels (1976) bases his critique of Monod's *Chance and Necessity* (1972) on the lack of precision concerning the definition of a random combination. For instance, consider the length of a random chord on a circle: What is the probability that it is bigger than the side of the inscribed equilateral triangle? Initially one might think there would be a definite answer to this question.

Schoffeniels (1976) gives three proofs showing that the probability is 1/3, 1/2 or 1/4, depending on how the chords are produced. Many other answers are possible. To be precise, we have to specify exactly the mechanism under which the chords are generated. This amounts to knowing the details of their probability distribution. Unless the context indicates otherwise, we may assume in this chapter that all random numbers are uniformly distributed over the range (0,1). That is, the probability density function by which the numbers are generated is rectangular over this range. Sometimes we will also consider random positive integers over a range delimited by the maximum integer that can be contained in a computer word.

There is a general feeling that randomness is associated with a lack of control, a lack of intention, a lack of order, or a lack of information about an occurrence. Not knowing precisely the mechanism, or not being able to account for the phenomenon, and thus preventing its exact repetition, also seem to involve randomness. The word definitely has negative connotations. The etymology of 'random' refers to impetuosity and doing things at a gallop.

The sequence of integers formed by the decimal expansion of $\pi$ seems quite random until we become aware of their deterministic origin. In fact, the first 16 million digits of $\pi$ have passed all the tests of randomness that have been applied to them (Wells, 1986). If we see a series expansion of $\pi$ in terms of fractions, we find it easy to guess the next term because we can readily recognise the pattern from the preceding fractions, but having to guess what is the next decimal digit would be very difficult. Knowing the mechanism seems to destroy the illusion of randomness (Hofstadter, 1980). Randomness as a hidden mechanism also forms the basis of encryption, where discovery of the encrypting mechanism would enable a third party to intervene in secret communications between two others (Hodges, 1985), which would otherwise seem incomprehensible.

It is difficult to come to an adequate definition of randomness. One attempt (Chaitin, 1975) links randomness with the concepts of computability and information, based on Kolmogorov's (1968 a,b) approach to the definition of information. The randomness of a sequence of numbers is defined as the length of the shortest program of a universal Turing machine which can produce the sequence. An orderly, repetitive sequence would need only a short program, whereas a more disordered sequence would require a more intricate, and therefore longer, program. There will come a point when the program approaches the length of the sequence itself. When this occurs, the sequence is random.

Creativity and randomness have traditionally been regarded as opposing forces. In Sumerian myth, the world was created by emerging out of aquatic chaos, presumably expressing Man's increasing control over Nature, particularly in taming the flood for irrigation. Yet in other cases, artistic creativity and randomness have gone hand in hand. Shuffling words, pictures or ideas in uncontrolled ways can inject new significance into what were previously disparate concepts. These methods connecting randomness with inspiration have been extensively made use of in Zen Buddhism and in European Surrealism (Hofstadter, 1980).

Randomness is often made use of in games to ensure a 'fair' distribution of items among players. Shuffling cards, taking numbers from a hat, rolling dice, etc., are randomising procedures employed to ensure that a game is not just a matter of pure skill, but to encourage the less-skilled to take on the more-experienced players. True estimation of skill will come about by long-term averaging, which reduces the effect of randomness. Alternatively, we may arrange the play so that everyone is faced with the same random selection, as in duplicate Bridge, which can be seen as an attempt to apply experimental design to negate the effect of the 'fall of cards'. Spinning coins, mixing dominoes or mah-jong tiles are normally difficult to control; yet if by sleight of hand some control can be exercised, there is the opportunity for deceit.

Chess is a game with no randomness whatsoever, apart from deciding who moves first. As the game progresses, each move reveals a further extensive set of possible moves. Strategic moves may allow or impede later moves, or tempt or rebuff moves of the opponent. Skill in chess presumably lies in the apperception of these moves and making an intelligent selection from them. Chess is exciting despite chance playing virtually no role, and has been at the centre of discussions about artificial intelligence. Snooker appears to have more randomness than chess, but under controlled conditions can also be quite deterministic. It can be seen as a 'robotic' analogue to chess (Kempf, 1983), the randomness being in inverse proportion to the precision of mechanical control.

The role of chance as an explanatory factor in physics started with statistical mechanics and Brownian motion, towards the end of the nineteenth century. Before this time, it was thought that if only the position and momentum of each atom could be known, then all phenomena — past and future — could in principle be determined. Statistical mechanics freed science from this deterministic strait-jacket. The concept of probability, originating from mathematical studies of gambling, was seen as a convenient way of modelling large numbers of uncoordinated actions. Thus, in Boltzmann's statistical formulation of the second law of thermodynamics, the inevitability of entropy to increase is only a statistical tendency, but ordered states of Nature being overwhelmingly outnumbered by disordered states, the tendency becomes almost a certainty. With the subsequent rise of quantum mechanics and information theory, determinism was no longer necessary for scientific 'respectability'.

Whether chance really exists in Nature at a fundamental level has been a bone of contention ever since it was asserted, in the Copenhagen interpretation of quantum mechanics. Einstein refused to accept it, with his affirmation 'God does not play dice', which was not an admonition against gambling, but an expression of doubt that randomness could have a primary existence in Nature. He preferred to think that observed random behaviour was a result of as yet undiscovered 'hidden variables'. Recent experiments have tended to confirm the Copenhagen interpretation (Gribbin, 1984). The kind of universe which is consistent with this interpretation is discussed by Bohm (1980).

Many features of designed systems rely on the fact that demands on their services will not be synchronised. For instance, banking relies on it being extremely unlikely that all depositors will one morning arrive on the doorstep and demand their balances in cash. Any system offering a service (booking system, supplier of goods) could not contemplate all potential customers turning up at the same moment. Although not impossible, its occurrence may be dismissed as very unlikely by the same wave of the hand which effectively proves the second law of thermodynamics.

Let us see how we make use of randomness in modelling designed systems. Consider, for example, the familiar problem for a supermarket manager of trying to decide how many check-out operators to employ to ensure a certain standard of service. The manager knows from experience, or perhaps by performing a survey, the statistical distribution of service times and of customer inter-arrival times that typify the workload. As we saw in Chapter 2, a model is identified by a boundary, which can be both external and internal. In this case the external boundary is clearly the point where the customers enter the supermarket. It is of no interest to the supermarket manager to investigate the reasons *why* the inter-arrival distribution is as it is, so he will not concern himself with simulating the customers' outside activities.

Similarly he will not be too concerned with the make-up of the set of goods purchased, as his survey has given the details of the average and general spread of service times which typify the system. This lack of concern delimits the extent of the model to an internal boundary. Nevertheless, not caring about some factors deemed to be beyond the boundary does not mean necessarily that they should be constant, or be related to one another. If short inter-arrival times were more often associated with long service times then this would show itself as a bias, and devalue the results. Of course, if the simulationist has reason to believe that such a correlation between two factors really exists, then including a bias in the model is justified. The main reason why we need a controlled source of randomness is to *simulate indeterminacy* at the model boundaries.

As modellers, simulationists can be quite pragmatic: no matter whether chance is real or illusory, if we can analyse real phenomena by synthesizing similar behaviour based on plausible assumptions, then we may run experiments to decide between different amendments of reality. This is sufficient for our purposes.

Every ancient civilisation had its own favourite method to invoke the Hidden World. Examination of animal's entrails (especially the liver), cracks formed when hot needles pierce bones (especially ox scapulae, or tortoises' carapaces), portentous flights of birds and delirious ravings of Delphic priestesses chewing laurel were just a few mechanisms. Of course, such effects require a priest or other official to mediate before a proper interpretation of the evidence can take place (Fig. 5.1).

By the time of the early European Renaissance, a more mechanical sophistication was in evidence. In the pages of Lull's *Ars Magna*, a primitive

Fig. 5.1 — The need for an intermediary to interpret random phenomena: at Delphi the ravings of the Pythia were interpreted by the priest (Finlay, 1970); in simulation, random phenomena are used to activate the components of the model, which are then interpreted by the simulationist.

logic machine was incorporated, which employed geometrical diagrams for the purpose of discovering logical truths. Segmented inscribed discs were rotated to suggest new interrelationships between concepts (Gardner, 1982). The indeterminancy was seen as an opportunity for divine intervention. The idea became popular and later developed into a precursor of calculating machines, and thus of the modern digital computer.

Simulations, being partial representations of reality, cannot be indefinitely large and powerful, and cannot be expected to account for everything; their scope is constrained between internal and external boundaries. A simulationist, in pursuing his Black Art, therefore needs to involve some kind of non-determinism in his model. In the remainder of the chapter, we investigate methods for doing this, and how to interpret the results.

## 5.2   GENERATION OF RANDOM NUMBERS

### 5.2.1   Uniform random numbers

Because the concept of chance behaviour is quite alien to a precise, deterministic machine like a computer, to obtain random effects we have to find a method of generating numbers which *simulate* the properties of randomness. Since any procedure specified by an algorithm in a computer is scrupulously followed in execution, anything happening really randomly in a computer would be regarded as a fault. If ever the same operation, executed under the same conditions, were to provide a different result, then something would be seriously amiss.

In order to simulate randomness, we must specify a deterministic procedure and ensure that the resultant sequence of numbers has sufficient characteristics of a truly random sequence. In allowing this apparently random number (pseudo-random number) to pass as really random, we gain an advantage in that our sequence of random numbers is *repeatable*. This is of great benefit in debugging (verifying) the simulation program, because it means that we can repeatedly submit the test simulation to the same conditions of 'randomness'.

Early computers saw fit to provide a genuine source of randomness, such as the least significant bits of a clock counter, but such devices are not as convenient as a pseudo-random source. Only when non-predictability is essential are true random-number generators useful. The ERNIE generator used in the British Premium Bonds lottery is based on counting the particles emanating from the decay of a radioactive source. Any deterministic method would obviously be quite unsuitable for this type of application.

Informally, the idea of randomness is associated with doing something in a non-determined way. In shaking dice we must not be allowed to simply turn the dice over; they must be turned over many times, and in an uncontrolled manner; similarly with spinning a coin. But with computers, ultimately nothing can be left to chance. There is a general feeling that not knowing precisely the mechanism of re-assembling the sequence, and thus preventing its exact repetition, somehow ensures randomness. Knowing the mechanism destroys the illusion.

Based on this observation, what would happen if we took a completely crazy algorithm and observed its effect? If we write down an algorithm which takes a number and performs crazy things to it (like swapping round and inverting parts of its bit-pattern, taking its log, etc.) so that anyone would think it would bear no resemblance whatever to the original, then we might arrive at a random-number generator. We are certainly familiar with the reverse situation, where a weird number results from something crazy happening as a result of a programming error, such as treating a procedure reference as a real number, for instance. Knuth (1981) (Algorithm K) gives an example of a crazy algorithm, but the successive values it produces soon degenerate into a fixed number, or very short-length cycles.

The algorithm basically consists of a sequence of 10 steps. We start with a 10-digit decimal number $X$. The starting number's first two significant digits are taken. The first determines the number of passes to be executed through the remainder of the algorithm; the second determines the initial step. The remaining steps are transformations such as: replace $X$ by the middle of $X^2$; interchange upper and lower halves; decrease non-zero digits by one, etc. The algorithm was designed so that an experienced programmer would find the machine code of the program to be quite incomprehensible.

Given the contortions that the initial number $X$ is put through, it seems quite plausible that the algorithm could supply an infinite sequence of random numbers. But the first time it ran it soon converged onto the number: 6065038420, which transforms into itself! Other initial numbers lead to cycles of relatively short length. Knuth (1981) draws the moral that random numbers should not be generated by a 'random' method: some theory should be used.

A simple, but not very effective, procedure for generating random numbers is the mid-square method, originally due to von Neumann. Each new $m$-digit number is produced by taking the middle $m$ digits of a number obtained by squaring the previous number. This very simple method, which appears scrambled and has no conceivable physical significance, is unfortunately unsuitable because it soon becomes degenerate. For instance, for decimal integers in the range 0000–9999, if the central digits of the square of the previous number are 2500, then the next number is 06250000, with the same central digits. Further, its stochastic properties are difficult to analyse.

A *congruential method* uses a recurrence relation for the generation of a sequence of uniformly distributed numbers. The *linear congruential* method, due to Lehmer, is of the general form

$$X[i+1] = (aX[i]+c) \bmod m.$$

where the **mod** operator

$$h = i \bmod j$$

is defined as the remainder when $i$ is divided by $j$, for integers $h$, $i$ and $j$, i.e. there is an integer $k$ such that

$$i = kj+h, \quad \text{or} \quad i-h = kj.$$

We say '$h$ is congruent to $i$ modulo $j$'. Integer $j$ is called the *modulus*, and $h$ is the *residue*.

Any sequence of numbers is thus determined by the choice of starting number or *seed* $X[0]$, constant *multiplier a*, and *modulus m*. The proper choice of the modulus is usually determined by the size of the word of the computer on which the algorithm is to be run. For a computer operating with binary arithmetic, the most natural choice is $m=2^b$, where $b$ is the number of bits in the computer's representation of an integer. Usually the computer hardware defines $m$, and therefore $b$, for the user. So a good random generator on a particular machine is defined by a judicious choice of seed $X[0]$ and multiplier $a$.

There are some rules to allow a choice of $a$, $c$ and $m$ so that the sequence has the largest possible cycle (Hull and Dobell, 1962):

A linear congruential generator will have period $m$ if and only if:

(a)   $c$ and $m$ have no common factor greater than 1;
(b)   every prime factor of $m$ is also a prime factor of $a-1$;
(c)   if 4 is a factor of $m$ then it is also a factor of $a-1$.

Naturally, if the period is of length $m$ then all the integers in $[0, m-1]$ must appear in the sequence, i.e. the sequence will be a permutation of $[0, m-1]$. Different choices of $X[0]$ will generate different permutations.

We said that it is convenient to take $m$ of the form $2^b$, where $b$ is the number of bits (excluding the sign) in the machine representation of an integer. This is because the evaluation of the *modulo* operation is then simply performed, by extracting the $b$ least-significant bits and just letting the high-order bits 'fall off the top' of the word. But this convenience is bought at a price: the bottom $d$ digits will have a period of not longer than $2^d$. In particular the last bit will always be 0, always 1 or alternately 0 and 1. However, the effect of this is usually negligible if we are considering numbers in the range $(0,1)$, obtained from dividing each $X$ by the modulus $m$.

For a maximum-period generator, we can relate the absence of undesirable properties to its *potency*, which is defined as the least integer $s$ such that $m$ is a factor of $(a-1)^s$. That such an integer exists is ensured by the rules for a maximum-period generator. Knuth (1981) argues that a potency of at least 5 is required. For $m=2^b$, the maximum potency is $b/2$ if $b$ is even, $(b+1)/2$ if it is odd, which is achieved by taking a multiplier of the form

$$a = 8k+5, \quad \text{for } k = 0,1,2, \ldots .$$

Another kind of congruential generator is the *multiplicative congruential* generator, of the form

$$X[i+1] = aX[i] \bmod m$$

where $a$ and $m$ are positive integers. They can be regarded as special cases of the linear congruential generator by taking $c$ equal to zero. Their attraction lies in being simpler, and hence possibly faster, avoiding one addition per generation. For certain values of $m$, i.e. for $m$ prime, or a product of low powers of primes, they have better statistical properties than linear generators. But, with $c=0$, condition (a) does not hold, so the generator cannot have period $m$. When $m=2^b$, the maximum period is only $2^{b-2}$, effectively discarding three-quarters of the possible range. To achieve even this range, the multiplier must be of the form

$$a = 8k+3, \quad \text{or } a=8k+5, \quad \text{for } k= 0,1,2,\ldots$$

and the seed $X[0]$ must be odd. All numbers in the sequence will then be odd.

The commonest random-number generators in use are the congruential ones. This is so despite their well-established tendency to generate dependencies between successive members of a sequence. We close this section with a brief look at some other types of random-number generators that have been suggested.

The *Fibonacci* (or *additive congruential*) generator has the general form

$$X[i+1] = (X[i]+X[i-1]) \bmod m$$

where each new number is the sum of the two previous ones, reduced modulo $m$. It is so called because of its similarity to the Fibonacci series. This simple method is seriously flawed by having extensive serial correlation, but more recent approaches, which take lagged members of the series combined by some binary operation, are effective.

Tausworthe (1965) has suggested a bit-stream generator based on repeatedly applying a linear transformation to generate a sequence of binary vectors, which can then be interpreted as a sequence of uniform random integers. This is also called a *shift-register* generator. All arithmetic is modulo 2, and addition is performed by the exclusive-OR operation. The linear transformation is usually chosen so that it can be performed by simple computer operations. Any non-zero integer may serve as a seed, and effective linear transformations to ensure maximum period are easily characterised. More details are given in Bratley *et al.* (1983).

Marsaglia (1985) gives a simple proof that a *combination* of two random-number sequences from different generators, using some kind of binary operation, will produce a sequence which should be more random than either of the originals. Not only should the combined sequence be more uniform, but also more independent. The binary operation need only have the property that its operation table forms a Latin square. One would therefore expect a combination, say of a lagged Fibonacci with a Tausworthe, to have very good properties.

Another source of apparent randomness, which perhaps may be used for generating random-number sequences, is *chaotic generators*. Clark (1985) considers one specific kind of first-order recurrence relation:

$$X[i+1] = 2k(0.5-|X[i]-0.5|),$$

the right-hand side being, for

$$0 \leq X[i] \leq 1,$$

an isoceles triangle with height $k$ and base 1, when $X[i+1]$ is plotted against $X[i]$. If

$$0.5 < k < 1 \quad \text{and} \quad 2k(1-k) \leq X[0] \leq k$$

then all members of the sequence

$$\{X[0], X[1], X[2], \ldots\}$$

will also lie in the interval

$$2k(1-k) \leq X[0] \leq k.$$

The sequence exhibits remarkable properties, including there being an uncountable number of seeds $X[0]$ for which the sequence is *not* asymptotically periodic, but *chaotic*. This property can be used to generate random numbers. If $k$ is taken to be $(1-e)$ for small positive $e$, the interval is practically $(0,1)$. For the particular values used by Clark, a limit cycle of over 3 million values was entered.

Applying Kolmogorov–Chaitin definition of randomness, the shortness of the programs for chaotic and congruential random-number generators would suggest that their output sequences could not be very random, despite the ability of at least some generated sequences to pass statistical tests of randomness. For supporters of the definition (Yakowitz, 1977), this observation merely demonstrates the weakness of independence tests.

## 5.2.2   Testing for randomness

Since there is no physical definition of randomness, any tests of the random character of pseudo-random numbers put forward will, in a circular way, define what randomness is, or what we would like it to be. Basically, there are two properties which we would like random numbers to possess:

- they should be evenly distributed over the range;
- each should be independent of the other.

Here we are concerned with uniform random numbers; in the next section we will see how to apply a transformation to a uniformly distributed random variable to obtain a sample from a non-uniform distribution. The first property ensures uniformity, i.e. that there are no gaps in the range and no clumping nor bias towards any particular sub-range. The second property, which is more difficult to test for, refers to the sequential relationship between successive values; ideally we would like no correlation between them.

Random numbers generated by the congruential methods will eventually form a cyclical set, with a maximum period bounded by the modulus. Normally the modulus is sufficiently large for the generator to supply enough random values for most practical purposes. But what if the period is shorter? There is a test, due to Floyd (Yakowitz, 1977), for the cycling of a sequence, which computes for any given generator, the quantities $X[n]$ and $X[2n]$ using stored values of $X[n-1]$ and $X[2n-2]$. The test terminates when

$$X[2n] = X[n],$$

and termination is guaranteed to be within the first cycle, so that all the numbers generated up to this point are cycle-free.

The generated numbers are essentially integers in a bounded range. We may examine the randomness of the integers themselves, or as real numbers in the range (0,1) by dividing the generated integer by the modulus. The following tests select a particular feature of the generated numbers (Mac-Laren and Marsaglia, 1965) and apply a comparison with the ideal situation based on either the chi-squared or the Kolmogorov–Smirnov significance test. Full details of testing are given in Mitrani (1982) or Bratley *et al.* (1983).

The *frequency test* looks at the uniformity, or evenness, of distribution of the generated sequence. The range is divided into equal intervals and a count taken of the numbers falling in each interval. The numbers are compared with the expected frequency which is proportional to the interval width. Other tests attempt to detect deviations from independence, discovering hidden patterns in the sequence.

The frequency test may be extended to look at the uniformity of consecutive *pairs* of numbers on the unit square in the *serial test*. Each non-overlapping pair of numbers is regarded as the coordinates of a point in the unit square. The (0,1)×(0,1) square is divided into equal-area sub-squares and the counts of pairs falling in each sub-square are accumulated and compared with expected values. The procedure can be extended to consecutive triples falling in a sub-cube of the unit cube, and so on to consider $n$-tuples in $n$-dimensional space. This method would be the severest test of the sequential properties of a supposed random sequence, but in practice it is too time-consuming for high dimensions because of the large number of samples required.

Congruential generators have been found to be particularly weak as regards sequential independence. They remain popular despite their well-established tendency for $n$-tuples of generated sequences to clump together on parallel hyperplanes (Marsaglia, 1968). For instance, using any multiplicative generator with a modulus $2^{32}$, fewer than 41 hyperplanes will contain all 10-tuples.

For simpler tests of long-range independence, we may look at particular characteristics, and compare the observed frequency of their occurrences with theoretical ideal behaviour. If we take, for each non-overlapping $k$-tuple, the largest member, $X[i]$, then if the sequence is independent, the cumulative distribution function of $X[i]$ is $x^k$. Thus we may test for a significant deviation of the frequency counts from the expected frequency. This is called the *maximum-of-k* test.

For the *poker test*, we first convert the numbers to integers in a certain range, typically [1,13]. Then we treat each non-overlapping 5-tuple as a poker hand, and compare the frequencies of different 'hands', for instance the configurations: all different, one pair, two pairs, three of a kind, etc., with their expected values.

*Autocorrelation tests* test the correlation between $X[n]$ and $X[n+k]$ where $k$ is the *lag*. For convenience, we can consider numbers to be in the range $(-0.5, +0.5)$, by taking

$$Y[i] = X[i] - \overline{X}.$$

Then if we form the autocorrelation sum for $Y$:

$$R(k) = \frac{1}{N-k+1} \sum_{j=0}^{N-k} Y[j]Y[j+k]$$

we would expect $R(k)$ to be zero for $k=1,2,3, \ldots$, and one-twelfth for $k=0$.

While the uniformity (or equidistribution) properties may be accurately assessed, no independence test exists which is completely adequate (Yakowitz, 1977). Despite this, random-number generators are widely used, and the good ones seem to have properties which are not troublesome, although there is always a nagging doubt. One could also argue, in connection with random sequences, whether it is more random to allow numbers to repeat before the period is out. Of course, having generators in which each generation is a strict function of the previous prevents this. But consider a set of $m$ sequences of full periods of $n$ numbers: if the whole set were then shuffled, would the result be regarded as being *more* random?

### 5.2.3   Random sampling from probability distributions

Often the random numbers needed in a simulation are distributed non-uniformly, and a means must be found to generate random samples from a given probability distribution. Which distribution should be used? This question may be partially answered by an appeal to the properties of the respective distributions. If there is no distribution which possesses the required properties, then the simulationist can specify a histogram of empirical measurements. The methodological aspects of this question will be considered more closely in the next section; here we look at how to generate samples from distributions other than uniform $(0,1)$.

Given $u$, a sample from a uniform distribution over the interval $(0,1)$, the simple linear transformation

$$v = (b-a) u + a$$

gives $v$ as a sample from a uniform distribution over $(a, b)$. For non-uniform distributions, other methods are required.

The commonest method is *inversion*, which uses the cumulative probability density function (cdf) $F(x)$ of the particular distribution from which a sample is required. Since any probability density function (pdf) $f(x)$ is necessarily non-negative and integrates to unity over its range, the corresponding cdf will be monotically increasing and will range over the interval $(0,1)$ (Fig. 5.2). A random sample from $f(x)$ may then be taken by first sampling a uniform random number $u$ from $(0,1)$ and using the inverse function $F^{-1}(u)$ to obtain a value on the abscissa. In order to perform the inversion operation in a program, we require to have the inverse cdf in a closed form, i.e. it must be expressible in terms of a function. For distributions for which a closed inverse function is not available, such as the normal distribution, we can use a numerical inversion technique.

Alternatively, we may approximate the cumulative distribution by a histogram, i.e. a table of cumulative frequency counts, and apply the inversion to obtain a sample from a continuous pdf $f(x)$ (Fig. 5.3). The same technique may be used to obtain samples from histograms of empirical data. Both discrete and continuous pdfs may be sampled. In the continuous case we may add refinements to more accurately inversely interpolate the cdf, but in practice this is rarely carried beyond linear interpolation. In some simulation languages, e.g. GPSS, this is the only means available to sample from a continuous function.

If the pdf is bounded, we may use a *rejection* technique. We can imagine the procedure in terms of a graphical representation of $f(x)$, as in Fig. 5.4. A rectangle of width $b-a$ and height $h$ is constructed around the pdf and two uniform random numbers are taken so as to generate a random point in the rectangle. It is best to choose $h$ as the maximum value of $f(x)$ over the range

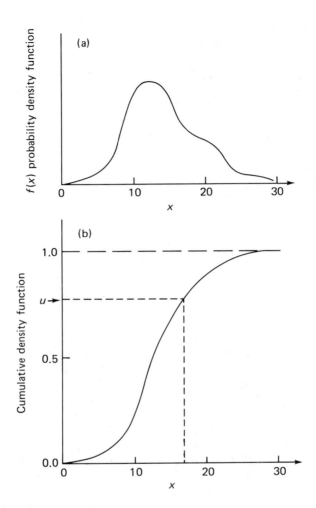

Fig. 5.2 — (a) A probability density function $f(x)$ over the range $(0, 30)$. The integral under the curve is one. (b) The cumulative density function of $f(x)$, $F(x)$ over the same range. To obtain a random sample from $f(x)$, a sample $u$ from a uniform $(0,1)$ distribution is taken, which is used to inversely interpolate a value in the $(0,30)$ range using the *cumulative density function*. In this example, $u=0.780$, giving a sampled $x$-value of 17.1.

of interest. Formally, if $u$ and $v$ are random samples from a uniform distribution on $(0,1)$, we calculate

$$x = a+u(b-a) \quad \text{and} \quad y = vh.$$

Then we accept $x$ if

$$y < f(x) \ ;$$

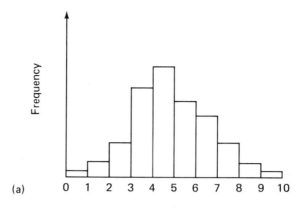

(a)

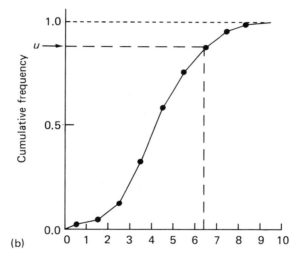

(b)

Fig. 5.3 — (a) A histogram showing the distribution of frequencies of occurrence in ten classes: 0–1, 1–2, ..., 9–10. (b) The corresponding cumulative frequency polygon. Each frequency is plotted at the midpoint of its class and the points are joined by straight lines. To obtain a sample from the frequency distribution, a sample, $u$, is first taken from a uniform $(0,1)$ distribution, which is used to inversely interpolate an $x$-value, using the cumulative frequency polygon. In this example, $u=0.880$, giving a sampled $x$-value of 6.37.

if not, a value for $x$ is retaken from a newly sampled $u$. The accepted values of $x$ will be distributed according to the pdf $f$. A comprehensive review of further sampling methods is given in Bratley *et al.* (1983).

For the exponential distribution, with pdf

$$f(x) = \lambda \exp(-\lambda x), \quad \text{for } x \geqslant 0,$$

and cdf

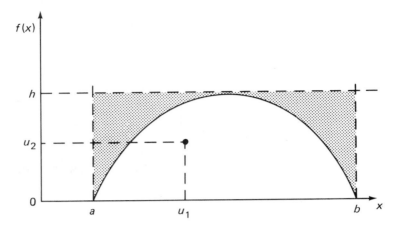

Fig. 5.4 — Rejection sampling. The sampling probability density function $f(x)$ is bounded by $(a, b)$ along the x-axis and by $(0, h)$ along the y-axis, where $h$=max $\{f(x) \mid a{\leqslant}x{\leqslant}b\}$. Two samples are required: one, $u_1$, from a uniform $(a, b)$ distribution, and the other, $u_2$, from a uniform $(0, h)$ distribution. If $u_2{>}f(u_1)$, the samples are rejected and new samples are taken. Otherwise, $u_1$ is accepted as a random sample from the probability distribution $f(x)$. Points $(u_1, u_2)$ which fall in the shaded area are rejected.

$$F(x) = 1{-}\exp(-\lambda x), \quad \text{for } x \geqslant 0,$$

an application of inversion gives

$$X = -1/\lambda \log (1{-}u)$$

as a generator of exponentially distributed $X$, where $u$ is distributed uniformly on $(0,1)$.

The Box and Muller (1958) method generates two independent standard normal values $x$ and $y$ from two samples $u[1]$ and $u[2]$ from the uniform $(0,1)$ distribution:

$$x = \cos (2\pi u[1]) \sqrt{(-2 \log (u[2]))}$$
$$y = \sin (2\pi u[1]) \sqrt{(-2 \log (u[2]))}.$$

However, Bratley *et al.* (1983) observe that successive pairs of normal values generated by the Box–Muller method are not independent when the uniform samples are generated from successive values obtained from a linear congruential method; in that case the successive values of $x$ and $y$ will fall on a spiral.

Details of methods for sampling from other distributions may be found elsewhere, e.g. from Shannon (1975), Fishman (1978) or Bratley *et al.* (1983).

### 5.2.4 Monte Carlo methods

Monte Carlo problems were some of the first applications of programming computers, especially in connection with calculations for the atomic bomb (Metropolis and Ulam, 1949). Apparently the method was first suggested to von Neumann by Ulam (Goldstine, 1972). By its nature it is an inelegant, brute-force method, and therefore disliked, quite naturally, by mathematicians. However, its continued usefulness in many areas of physical science has proved its worth. According to Heerman (1986) the field of 'simulational physics' is rapidly expanding.

It is worthwhile mentioning here the distinction between discrete-event simulation and Monte Carlo programming. Although the two make use of pseudo-random numbers, their respective goals are quite different. In simulation we are concerned with adequately representing a system of parallel interacting processes, in which randomness plays a role in increasing the realism of the model. Monte Carlo programming, on the other hand, is best regarded as a technique of numerical analysis, primarily a method of integration. Rivett (1980) strongly insists on the distancing of simulation from Monte Carlo programming.

The basic Monte Carlo approach is similar to the rejection technique discussed above. We keep a score of the number of accepted $x$-values to the total number of $x$-values sampled. Clearly, their ratio will estimate, in a rather haphazard fashion, the integral of $f(x)$ over $(a,b)$. Obviously this is a rather unsatisfactory method, resembling throwing darts randomly at a board on which a graph of the function $f(x)$ is displayed, and counting the 'hits', the darts which fall under the curve. It is literally a 'hit or miss' method. It is only useful when random samples are available in rapid succession. The method can be speeded up by the use of control variables (Yakowitz, 1977).

Yakowitz (1977) also compares the use of Monte Carlo with more classical approaches to integration. He reports that Gaussian quadrature can be quite competitive with Monte Carlo integrations for well-behaved functions involving three to six variables. But if the functions are discontinuous, or even fail to possess continuous first partial derivatives, Monte Carlo methods are often superior.

There is an eighteenth century procedure for experimentally determining $\pi$ suggested by Buffon (1707–1788), involving a 'randomising' procedure, which can be regarded as an early precursor of modern Monte Carlo methods. A needle of length $L$ is thrown 'randomly' onto a board ruled with straight lines, distance $L$ apart. It can be shown that the probability of a needle intersecting a line is $2/\pi$. For the problem to be stated in meaningful terms, we must assume that the needle is thrown such that its centre is uniformly distributed over $(-L/2, L/2)$ from the nearest parallel, and its orientation to the lines is uniformly distributed on $(0, \pi/2)$.

We will not dwell further on Monte Carlo methods, for the reasons already stated. Having described how to generate randomness in a program, we turn now to how best to use it in the methodology of construction and experimentation with the simulation.

## 5.3 RUNNING SIMULATION EXPERIMENTS

In this section, we take an overall look at the issues raised by the methodology of experimentation with systems containing random components. The main concern is how the simulationist can gauge the significance of his findings with respect to sources of randomness in the model; in other words, how can he judge the extent to which his findings are due to chance? Furthermore, how can the precision of the findings be improved upon? Basically, there are two problems: the strategic problem of designing the experiments so that the desired information can be obtained, and the tactical problem of conducting the experiment efficiently.

This is a vast topic, engulfing a great part of statistics, and only a superficial account can be given here. Good introductions to the interface between simulation and statistics are given in Bratley *et al.* (1983), Fishman (1978), Law and Kelton (1982) and Mitrani (1982). A more advanced text is Kleijnen (1987). We present here only a brief look at the main outlines of the area, and reference to the above-quoted works is recommended for further details.

### 5.3.1 Selection of probability distributions
The choice of the probability distribution by which to model the durations of activities and the inter-arrival times of temporary entities is an important part of the model and deserves serious consideration. Probability distributions are often also implicated in assigning values to entity attributes. The real system, which is the source of most of the information about the simulation, should be intensively surveyed. Relevant measurements include the timing of occurrences, activity durations, arrivals, the occupancy of facilities, build-up of queues, etc. To get a total overview of the system, the survey should include a questionnaire and interviews with managers and other system controllers at all levels.

Observations on the real system form the basis of most of the quantitative information employed in the simulation. In the meantime, we should take the opportunity to observe any possible interaction between variables and look out for any significant fluctuations, such as peak periods or diurnal variation. Even if some of the collected data is redundant, it may be used to ensure consistency. Where the relevant part of the real system is only hypothetical, then we may have to make a guess as to a likely distribution.

When we have obtained our data, we may either introduce the empirical data directly into the simulation, or fit the data to a probability distribution and use it as a distribution to be sampled from, as described in the preceding section. Standard statistical techniques exist to fit empirical data to a theoretical distribution. Once a suitable distribution has been chosen, the fitting procedure will estimate values of the distribution's parameters. If raw data are to be used, then it is usually convenient to classify the information in the form of a histogram, where the range is suitably subdivided so that each class has several data points. Sampling from a histogram follows the same procedure as sampling from a discrete approximation to a continuous distribution (Fig. 5.3).

There are arguments for and against using empirical data directly. Most statisticians tend to think in terms of a hypothetical distribution which underlies the observations made at a particular occasion. Thus a distribution is 'truer' because it does not involve the particular circumstances which prevailed during the time the survey was undertaken.

Using a histogram of direct observations as a discrete sampling distribution has the unfortunate consequence that no samples will be generated which fall outside the extreme class boundaries, which could cause the simulation to ignore the possibility of outlying points. The omitted points could have put the system under extra stress which could cause the simulation to report the system more favourably than it should. Using a distribution, it is easier to generate values over an infinite range, if necessary. Often distributions are used because of their hypothetical nature: for instance, customer arrival is almost invariably modelled by a negative exponential distribution, because this corresponds to a uniform distribution of arrival events.

The choice of distributions is wide. Law and Kelton (1982) give a handy compendium of facts about eight common continuous distributions, together with graphs of their pdfs, and six discrete distributions, with their mass functions. Before attempting to fit data to a distribution, some thought should be given to the general *family* of distribution which is appropriate. There are various guidelines that can be used to select a family. For instance, we might just compare the general shape of the raw-data histogram with the graphs of probability density functions.

Another such guideline is an estimate of the *coefficient of variation*, which may be calculated by dividing the sample standard deviation by the sample mean. The value obtained may lead us to reject some distribution families. When the coefficient of variation cannot assist in indicating the family, other functions may be considered, such as coefficients of skewness and kurtosis.

Having chosen the family of distributions, the simulationist must choose the parameter values which define more closely the shape of the distribution..There are many ways of estimating the parameters of a distribution. The most popular estimators are *maximum-likelihood estimators*, which possess several desirable properties such as asymptotic normality. Further details are available in the statistical literature. Often it is worthwhile to check that the chosen distribution closely fits the empirical data obtained from fieldwork. This can be performed by goodness-of-fit tests, of which the chi-squared and Kolmogorov–Smirnov methods are widely used.

If for some reason there is little real activity on which to base observations, one possibility is to estimate a minimum time and a maximum time for the activity. In the absence of any other information, a uniform distribution can be assumed over this range. If it is thought that a uniform distribution would give undue emphasis to the extremes of the range, then a 'most likely' value within the range can be posited and a triangular distribution assumed.

### 5.3.2  Approximation

In the course of constructing the simulation, some factors operating in the real world are discarded, and, in compensation, various simplifications and approximations must be made. Validation problems arise because of the approximations that we are forced to make in any modelling process. Validation addresses the problem: to what extent does the behaviour of the simulation carry over to the hypothesised reality? Or, more succinctly, are the model results valid? Do they really mirror reality?

Probably the largest approximation made in most simulations concerns the *model boundary*, the demarcation between phenomena of interest in the system, and phenomena in which we express no interest. Frequently the model boundary coincides with a real boundary, such as when we decide to model the arrival of customers in a supermarket system by using an inter-arrival distribution. This amounts to saying that the arrivals occur randomly at a level rate, without undue clumping or synchrony with any other effect in the model. Often it is simpler to assume that components are independent of one another.

To model realistically the arrivals of customers would presumably involve a large socio-psychological system with complex hypotheses, and be largely irrelevant to the system of interest. It is of philosophical interest to consider whether all assumptions of probability distributions are of this type, i.e. whether they are relative approximations, or if some real-world phenomena are *really* random.

The homomorphic methodology of simulation encourages an *aggregation* approximation, where components tend to be lumped together and treated as one. The definition of a process as an archetype of an individual entity inevitably leads to entities tending to lose their uniqueness and becoming stereotyped. This is partly a convenience for modelling and partly a consequence of the simulationist wishing to see the problem in general terms. Aggregation is also performed in the course of recognising a pattern of activities. For any real system, one simplifying procedure is to abstract patterns of phenomena from reality and describe the abstraction in the simulation. Then connections can be made to real instances of the abstract archetype.

Aggregation is another example of the divide between template and instance: the process-entity relation is well known but the process itself can be an instance of a larger template. Many complex multi-stage processes can be structured in this way. For example, consider an assembly line in a manufacturing plant, or a lift system. One stage, expressed in general terms, can be the basis for the whole process. For a lift system, the process defining the lift would simply consist of the following segments:

- move to the next floor;
- let passengers out;
- let passengers in;

since essentially the same logic is repeated at every floor, with due regard to direction of travel, etc. The resulting process is also more flexible; a simulation of a particular lift system becomes, in a way, a general simulation of all lift systems, or at least a starting framewwork.

### 5.3.3  Verification

To *verify* a simulation primarily means debugging the simulation program and testing it for consistency, both internal consistency within the program, and its correspondence with respect to its prototype model. One of the roles of a simulation language is to steer the use of its concepts in a logically coherent fashion. For simulation-language designers, the needs of verification encourage the language to enable the expression of the program in terms similar to those used in the prototype. We will consider in Chapter 6 and Chapter 8 the ways in which model and program forms of language may converge.

As in any other type of programming language the influence of the language design is felt through the syntax checking carried out by the compiler. A source-program editor, attuned to the syntax of the host language, would speed up the programming stage considerably, and pre-empt discovery of many of the errors found normally at compile time. Once the program is free of syntax errors, a run-time debugging procedure should be undertaken, starting from quite simplistic assumptions, gradually building up the complexity of the simulation in stages.

Thus we start with simple numbers for activity durations, to enable comparison with hand-checked calculation, and continue stage-by-stage replacing the simple assumptions with more realistic data. Inserting and deleting test data would be easier the more separate the process and data specifications. A facility for enabling a run-time trace of selected events, or all events partaken by a selected transaction, would also be helpful.

Adopting good, structured-programming practice means that the components of the program should be able to stand alone as program modules, which facilitates their being tested individually. For instance, subprograms such as subroutines and functions interact with surrounding modules via explicit parameters. These tests can be regarded as first-stage verification. But we should note here that the standard structural methodology may not apply directly to a simulation program, especially when using a PI language. This is because the important and interesting phenomena in simulation are concentrated around the effects of the *interaction* between entities, so although each entity description is written as a module, it may not be able to stand alone, and its significant effects can only be appreciated in interaction with others.

As a second stage, the simulationist might want to consider replacing the random components with samples from the exponential distribution; sometimes this leads to a system for which the solution is known, from queueing theory. Further verification is possible by varying a single major model parameter at a time and checking if the results are 'sensible'. Bratley *et al.* (1983) suggest *stress testing* whereby the parameters are set to unusual values, and then the output is checked to see if the simulation 'blows up' in an understandable fashion. The kinds of stress they suggest is the operation of the system under loads greater than 100% or the replacement of distributions by the Cauchy distribution. They claim that bugs that would otherwise remain obscured often manifest themselves under stress.

### 5.3.4  Experimentation

The main problem of experimenting with systems comprising random components is that the results obtained will also have some random component. Any measurement, such as the average queue length, or average waiting time, will therefore be only an estimate, with an associated mean and variance. To improve the precision of simulation results, we can improve the statistical efficiency by *variance reduction*.

Many variance-reduction schemes have been proposed, although most of the work has been carried out in the field of Monte Carlo programming. We shall describe two main approaches to reducing variance in simulation. The first, that of using *common random numbers*, is applicable in the case where we are comparing the results of two simulations. This is a very common situation in experimentation, which usually occurs when comparing the same system under different values of a particular parameter. For instance we might be interested in running the same supermarket simulation, with a different number of servers in each case.

Common sense would suggest that each experiment should be conducted under identical conditions to ensure a fair comparison between parameter effects. In particular, the random numbers which schedule arrivals, determine shopping lists and check-out times, etc., should be exactly repeated for each condition. To do this requires some ability to repeat the same random-number streams from their initial seed values. Fair comparisons are also assisted if each point where random samples are drawn in the entity process makes use of a separate random-number generator, perhaps in the form of a single random-number sequence with starting values widely spaced along the generated sequence. These seeds should be associated with the activities for re-initialisation. In SIMSCRIPT there is a facility for ten random-number streams (Fishman, 1978).

Sometimes the factors to be compared between two simulation runs will impede a fair comparison. Not only must the same random numbers be used, but they must be used in analogous ways across the comparison. The particular structure of the simulated system can have an effect. For instance, if the discipline of a queue is the subject of the investigation, then it might be better to assign a service time to each customer entity as a parameter value on arrival, rather than at the start of serving which is more usual. Otherwise that particular service time might be associated with another customer in a comparison run. For a similar reason, rejection sampling should be avoided, as the values of the rejection parameters, and therefore the necessity to take one or more samples, may be based upon model parameters to be compared between two runs.

Use of common random numbers attempts to encourage *positive correlation* between observations across sample runs for different systems. Within a particular system, it is useful to be able to induce *negative correlation* between samples, in order to reduce sample variance.

Around any estimate of a probability distribution parameter, we have a sample variance. For instance, suppose we wish to generate service times from a distribution with a mean of 5. If we take, say, 50 samples consisting of samples from a uniform $(0,1)$ distribution followed by a suitable transformation, we might find that the sample average is 4.95, which is fairly near the desired mean, but maybe not close enough for the purpose of the experiment. If the next 50 samples are taken using the same 50 uniform random samples, but subtracted from 1, then any bias towards the low end of the range will be exactly compensated by an equal and opposite bias, and will bring the sample mean into line with the distribution mean. The second set of random samples is termed *antithetic* to the first. Of course, inducing correlation by use of antithetic variables is only justifiable if it is consistent with model fidelity. Antithetic variables were used to reduce the variance in the comparison of future-event scheduling algorithms in Chapter 3.

### 5.3.5  Validation

The larger issue of simulation *validation* concerns the closeness of the approximation to the real (hypothesised) system for the intended application. We are concerned with encouraging confidence in the findings of the simulation. Validation is essentially a statistical question because the simulation will almost always contain elements of randomness. Validity is difficult to define, as it requires an all-round view of the simulated system in the context of its purpose, its effects, even its interaction beyond the boundary, and the relationship of these factors with their complement in the real (and hypothetical) world. Consequently it is difficult to lay down strict rules about how it should be conducted. Necessarily, it will involve a human manager or operator to exercise a judgement. Computers have not yet shown themselves to be competent enough to wield this kind of power, though by marshalling the information they can lessen the burden on the human decision-maker.

The largest stumbling block for the theoretical acceptance of any model of a stochastic system is the assumption of *stationarity*. It is a minimal requirement for a characterisation of stochastic processes. For a simulation, stationarity essentially means that its statistical properties at any value of simulated time remain invariant over time. Another theoretical requirement is *ergodicity*. A stochastic process is ergodic when successive samplings from a single (long) run are consistent with samples from repeated runs. These properties give us theoretical support to infer behaviour from sample data (Fishman, 1978).

However, common experience tells us that many features of everyday life, in the political, economic, social and environmental spheres, are grossly non-stationary, and the notion of a steady-state system is often vacuous. A steady state assumes that there is a balance between opposing forces, but many interesting day-to-day activities are not of this form. The stock market can be seen as consisting of non-equilibrium speculative movements taking place in the short term, against a background of deeper effects of movements in the economy. But it is doubtful whether the assumption of a steady state in economic affairs is anything but a convenient fiction.

A steady-state result is held to be of more significance since the effect of initial conditions will have been smoothed over. But many management situations demand that the current state *should* be a significant factor in future development. Many a 'what-if' problem seeks the short-term implications of starting from the current configuration of the system. A system manager is not necessarily pursuing eternal truths.

Kleijnen (1978) goes so far as to state that steady-state behaviour is of no practical importance, and stresses the importance of start-up and end effects. Consequently we should be skeptical about extrapolations of steady-state results to real systems. An interesting taxonomy of models based on the *purpose* of the model, which enables the appropriate validation tool to be selected, is given by Lewandowski (1982).

There are many ways of estimating when a steady state has been reached. Shannon (1975) lists six published recommendations for estimating when a steady state has been achieved, but none of them has theoretical backing, and they are all basically subjective. One must be careful of not biasing the model towards the desired result by choosing initial conditions which seem to favour a steady state. Such a procedure may suggest itself as a means of shortening a simulation run by starting off in a state near the expected steady-state configuration. Unfortunately this can lead to a paradoxical situation (Conway, 1963), in which the expected result is assumed in the choice of the initial configuration.

### 5.3.6 Output analysis
One feature which distinguishes a simulation program from a conventional calculation is that generally there is no single result: what is sought is a general indication of the way a system would go if it were changed in one of a variety of ways. As computing environments become more sophisticated, then the end result of a simulation might become an extremely flexible computer model, which could apply a 'what-if' analysis at the whim of a manager. Even so, at some stage it is important to come up with good *measures of performance*. There are two basic questions that experimental design of a simulation will pose (Bratley *et al.*, 1983).

(a)  what are good estimates of performance measures?
(b)  what are good estimates of the goodness of these estimates?

There are no water-tight answers to these questions: often graphs of response of a system with respect to a particular controlled condition are more useful than dealing with means, and often the graphs are highly skewed. Using simulation to optimise a system, i.e. attempting to find the 'best' set of conditions under which a system will run, is thus possible, but is rather unsystematic compared with other techniques of Operational Research, such as linear programming. On the other hand, the richness of expression allowed in a simulation requires fewer assumptions, and allows more of the full complexity of the real system to be represented.

Bratley *et al.* (1983) discuss output analysis with respect to three issues:

(a)    finite-horizon versus steady-state performance;
(b)    fixed sample size versus sequential sampling;
(c)    absolute versus relative performance.

The activities of many systems naturally lend themselves to considering their performance in terms of a finite time limit. This is so despite frequent claims that simulationists are interested mostly in steady-state behaviour. For instance, a system may operate with daily peaks and troughs, such as electricity supply, or may be considered to be slack when a specific task has been completed, such as an airline check-in counter servicing a specific flight. Within such horizons, the systems may not reach a steady-state. Also the effect of the starting configuration is, and should be, reflected in the analysis.

Where a steady-state performance is required, we have to estimate an infinite-horizon performance from a finite-horizon sample. Here, the starting configuration should be irrelevant, which leads to problems of statistical analysis. In particular, the simulationist must decide when to terminate a particular run. Basically, we can either choose to terminate the run after a fixed number of observations of the statistic of interest, or specify a precision beforehand.

In the first case, we do not know whether the estimate is precise enough and in the second case we might find that the number of observations required is unacceptably high. A valid confidence interval can be placed around the statistic if sufficient observations have been made so that the central-limit theorem allows us to ignore any non-normality in the data, and, for the case of steady-state simulations, that lack of independence and the bias due to initial conditions have been overcome by some means.

The logical end of simulation is the successful implementation of a 'better' system, with respect to a specific purpose. To be consistent, the simulationist should also be concerned with the implementation of a new scheme, and re-test its effectiveness. In some cases, especially where a system is expected to respond to rapidly changing conditions, the methodological cycle might be required to start over again, perhaps with different decisions about what is relevant to the model.

There are several pitfalls in the estimation of people's attitudes to new systems. Shannon (1975) mentions the 'Hawthorne effect' where people will change their behaviour when they know they are being watched. This is a problem of both data collection and implementation. Other studies, mainly in an educational context, have shown that any new system will show an improvement, not necessarily because of any real superiority, but just because it is being implemented with enthusiasm. The provision of extra service can also serve to *stimulate* its use. Transport systems are a case in point. Any alleviation of a traffic bottleneck can also create travel opportunities, while measures of congestion still remain constant. New systems can alleviate, or degrade, an existing level of performance, simply because they are different.

While some remarks of this section may seem unduly negative, even to the extent of casting doubt on the statistical usefulness of simulation at all, we may be reminded of an adage of Tocher's that once a simulation program has been written, the large amount of attention to detail and logical checking of the real system that will have taken place will reveal such a host of mistakes, missed opportunities and inefficiencies, that the actual running of the program is no longer necessary! While such skepticism should not be taken too seriously, it remains true that there is a lot more to simulation than a statistical model.

# 6

# Simulation language

## 6.1 DIAGRAMMATIC REPRESENTATION

### 6.1.1 Attractions of diagrams

In our discussion of simulation strategy in Chapter 4 we touched on several means of describing simulations, without fully considering the pros and cons of the various available alternatives. As we saw, a description expressed as a program necessarily entails the adoption of one of the three main strategies, namely event-scheduling (ES), activity-scanning (AS) or process-interaction (PI). On the other hand, using a diagram to describe the same discrete-event system does not necessarily commit either the reader or the describer to any particular strategy. In this first section we will investigate further the diagrammatic approach to system representation, and look at some conventions and notations that have been put forward.

In general, it must be admitted that diagrams do possess distinct advantages over textural representations, whether in the form of programs written in a programming language, 'structured' English, or natural English. This point has been made clear in some recent empirical studies (Vessey and Weber, 1986). The old adage 'a picture is worth a thousand words' seems to hold true for simulation as in other technical disciplines. The fact that even when programming we tend to pursue devices such as indentation to structure our programs seems to indicate the additional power of two-dimensional pattern in making text more readable.

It is from their two-dimensionality that the diagrams gain over programs for expressing system phenomena. Two entities processing simultaneously can be represented by simply juxtaposing their respective diagrams. Interactions between entities can be shown by the joining and subsequent separation of process lines. As long as the interactions are fairly simple, a diagram will be concise, enabling the structure to be seen as a whole.

A program, on the other hand, forces the description of the system into a sequential form, imposing an artifical ordering on the system components, whatever strategy is adopted. We run the risk of introducing the possibility of a bias in the interactions where none actually exists. The strategy adopted by the programmer confines the reader's attention to one particular viewpoint, which may not suit his inclinations, and could lead to difficulties in interpretation. The one-dimensional, sequential nature of programs is an inherent property, going back to Turing's earliest conception of a computing device, reading from and writing to a linear tape.

In practice, programming languages too often tend to reflect the niceties of a particular machine rather than letting the programmer concentrate on modelling the system structure, which is the prime aim of simulation (Knuth, in Buxton (1968), p. 53). Although this impediment generally lessens as the level of the language rises, even with the highest-level languages it is often necessary to include statements in the program which intrude into the communication.

Diagrams, since they are not imperative statements, are free of implementational restrictions, have a fairly obvious syntax and semantics (at least for simple cases), enabling them to cross linguistic boundaries easily and to exhibit global aspects of interactions more readily. Various authors have advocated diagrams for representing systems whether as an aid to understanding the eventual simulation program, or as a methodological step between the reality and the program. It is hopelessly optimistic to expect a modeller to jump straight from an initial perception of the real system to a program in one leap; our ideas must first be roughly sketched, elaborating them gradually into modules before submitting them to the necessary rigour of a programming language. A diagram serves as a helpful bridge between our vague internal ideas and rigorous external programs.

Regrettably computer language has not progressed to the stage where it seems at all natural to people not involved with computers. Perhaps it never will, but even so, a diagram will always be more readily assimilated by nontechnical members of the project team who may not grasp the intricacies of a program; a diagram thus helps in the dissemination of ideas about the model, enabling some consensus about the degree of detail and realism to be attained. A diagram is also a convenient way of publishing the broad outline of a simulation without getting tangled in the syntactic peculiarities of a particular language, e.g. Carrie et al. (1985). But useful as diagrams are, it is unlikely that diagrams will ever supplant programs; they can present typological nightmares, and would be difficult to present to the computer as direct input.

### 6.1.2  Diagrammatic methodologies

Tocher introduced his 'wheel charts' (Tocher, 1964) in order to provide a diagrammatic means of expression for 'machine-based' simulations, as opposed to the 'flow-based' approach of GPSS. A wheel chart consists of a set of linked boxes representing a sequence of activities for each machine. Each cycle of linked activities corresponds to a 'machine' (or resource) processing cycle giving an overall 'interacting cogs' description of the system. Alternatively, if the activities of a machine do not necessarily proceed in a strict sequence, the wheel chart takes on a star-shape; see Fig. 6.1.

The purpose of the wheel chart was primarily to assist in the identification of bound and conditional activities which is necessary in order to

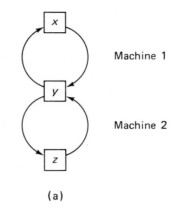

(a)

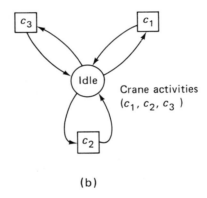

(b)

Fig. 6.1—Tocher's (1964) wheel charts. (a) Intersecting machine cycles. Activity $y$ is conditional, being on the intersection of two machine cycles, whereas $x$ and $z$ are bound to start immediately $y$ finishes. (b) Non-cyclic activities. The crane may perform any of the activities $\{c_1, c_2, c_3\}$ without regard to sequence. On completion of the activity, the crane returns to the idle state.

proceed with a GSP simulation program, composed in terms of B- and C-activity descriptions. If the number of arcs entering a node is one, then there can be no resource-limitation on that activity, so it is of type B. All others are of type C.

The activity-cycle idea was taken up by Clementson (1977) and adapted as a front-end, called CAPS, for the simulation language ECSL. The states of the system, represented by nodes in the diagram, are divided into two types:

- *activities*, which denote an active state performed by one or more cooperating entities for a period of time which can be determined at the start of the activity, and

- *queues*, which represent passive states, in which entities stay for an unspecified duration.

Conventionally, activities are represented by boxes, and queues by circles, and for any entity, the sequence of nodes on the cycle should alternate between queues and activities, even though this requirement may incur the insertion of a dummy queue. Hence they are often known as QAQA diagrams (Fig. 6.2). Let us consider a diagram of the following system:

An operator is required to set up a machine, which then operates automatically until one unit of production is formed. Then the machine stops until an operator is available to set up the next job.

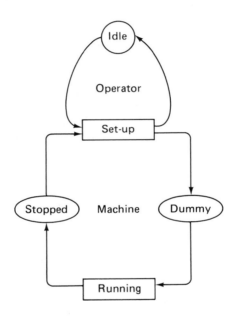

Fig. 6.2 — Clementson's (1977) QAQA activity-cycle diagram. Activities are denoted by rectangular boxes, queues by circles or ellipses. The dummy queue is necessary to separate the set-up activity, which involves the operator, from the running activity, which does not.

The diagram clearly shows the distinction between C-activities (called cooperative activities in CAPS, Clementson (1977)), which require the involvement of more than one entity, and B-activities, which do not. The diagram for the producer–container–consumer (P–C–C) model, introduced in Chapter 4, is of this type. Many other versions of activity-cycle diagrams have been proposed: for DRAFT (Mathewson, 1975, 1985) and for HOCUS (Poole and Szymankiewicz, 1977).

The general entity–resource interaction is diagrammatically represented by Birtwistle (1979) in a similar activity diagram, where the activities are represented by boxes, and resources in their idle state by circles, with numbers to represent the initial number of items present. Directed arcs connect resources with activities to indicate resource-limitation, and the number of arrow-heads on each arc represents the number of resources required for the activity.

At the same time as Tocher was developing wheel charts to represent the interaction of machine activities in an industrial context, Petri (1962), in a study of parallelism in computer systems, was putting forward superficially similar diagrams which are now known as Petri nets. Despite their similarity, the context in which each was developed was quite different, Petri nets being precise, mathematically rigorous statements, whereas Tocher's wheel charts, or activity-cycle diagrams as they became known, were informal, 'back-of-an-envelope'-style sketches with which to communicate with managers on the factory floor.

There are four basic components of a Petri net (see Fig. 6.3):

- transitions, represented by bars (or sometimes rectangular boxes);
- places, represented by circles;
- arcs, either from transitions to places or places to transitions;
- tokens, whose movement from place to place through the net gives expression to its dynamic properties.

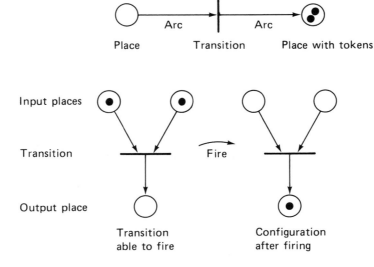

Fig. 6.3 — Petri-net elements, and an application of the firing rule.

For any sequence of directed arcs, the nodes are traversed in an alternating sequence of place and transition, so that for every transition we have a set of places immediately preceding it (the input places), and another set of places immediately succeeding it (the output places).

Activation of the net occurs by application of a *firing rule* to each transition. A transition fires if all its input places have tokens on them. In firing, the tokens are removed from the input places and transferred to the output places. For a simulation of a system of interacting parallel flows, we may identify the tokens with system entities, different types of entity being distinguished by colours. The paths taken through the net by tokens are also appropriately coloured. Thus a token positioned in a place represents an entity in a state, and the firing of transitions corresponds to an event occurrence. Note that there is no express rule to govern in what order the transitions should be considered for firing. As such the net can be used to express the dependency of occurrences on entities being in certain states and some features of parallel systems such as deadlock, contention, etc., but there is no representation of time, or that tokens may be delayed in the occupation of a place.

In order to incorporate the concept of an event occurring after a delay in time, we may consider the source of events, i.e. the FES, as a special 'place' which may be connected to any transition, the event-notices identifying the particular transition involved. We introduce the star symbol to indicate that

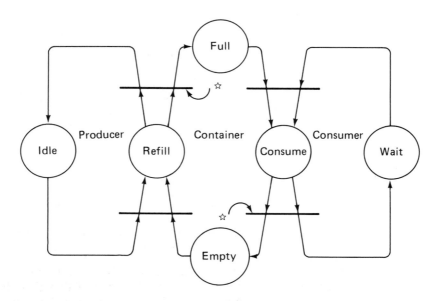

Fig. 6.4 — Petri net for the P–C–C model. The ☆ symbol indicates that a firing is further withheld until the occurrence of a temporal event.

B-activities: refill.end, consume.end
C-activities: refill.begin, consume.begin

a transition is time-dependent, as shown in Fig. 6.4, an augmented Petri-net diagram of the P–C–C model.

Although the augmented Petri net is superficially similar to the QAQA diagram, there are several immediate advantages. Firstly the distinction between events (transitions) and activities (places) is made plain, as well as the difference between B-activities (time-dependent, and therefore starred, transitions) and C-activities (transitions without stars). As we shall see in Chapter 7, the set of symbols can be further enlarged to cope with more complex phenomena.

One problem with most activity-cycle diagrams is their incapability of representing arrivals and departures of temporary entities to and from the system. Most often the departures are linked to the arrivals by a fictitious 'outside world' activity in order that the cyclicity of the diagram is maintained, e.g. Poole and Szymankiewicz (1977). While perhaps indicating that link between release and later re-allocation of memory space in dealing with dynamic data structures, this is a purely implementational matter, and should not intrude into the model. We may represent arrivals simply by a transition activated by a star, and departure by a place without an exit arc (see Fig. 6.5).

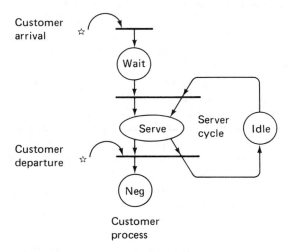

Fig. 6.5 — Petri net for customer–server interaction with arrival and departure conventions.

Our formulation of a Petri net to describe simulations is based on Reisig's formulation of a relation net (Reisig, 1985). Although both forms of diagrams have been used extensively in their respective fields, the connection between activity-cycle diagrams and Petri nets was first mentioned in print as late as 1981 (Törn, 1981). More recently, Petri nets have become

more widely appreciated in the control of manufacturing systems, e.g.
Favrel and Lee (1985).

To complement the diagrammatic forms already used by AS and PI
strategists, Schruben (1983) has put forward a graphical method to assist
simulation according to the ES strategy. Each node of the graph corresponds
to an event module and the directed arcs from each node indicate the event
modules which may be scheduled form the node. The diagram (Fig. 6.6) thus

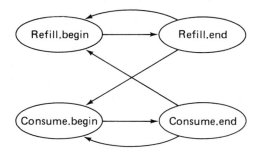

Fig. 6.6 — Schruben (1983) event graph for P–C–C model. The arcs indicate from
where event modules are scheduled.

addresses the representation of 'what schedules what' in the ES interpretation of the P–C–C model. Given the obscurity of the scheduling pattern
that can easily arise even in small models written according to the ES
strategy, together with the popularity of the ES strategy using SIMSCRIPT
in the USA, it is surprising that no one has come up with the idea before.

In terms of the second paradigm of simulation, a diagram such as the
augmented Petri net described above, concentrates on the sequence of
engagements and the interactions between different flows. Data-probes are
concerned with the passage of entities across transitions, and do not form
part of the model representation itself. Allocations, however, tend to
involve operations on entity-attribute values, and can be shown only with
detriment to the clarity of the diagram; they are best described separately, in
textual form.

The diagram is therefore not to be regarded as an isomorphic equivalent
to its corresponding simulation program, but a skeleton containing the
broad strokes of the flow and interaction pattern. In the absence of
allocations, the diagram should show the links between *possible* interactions
only; what will actually happen may depend on properties defined at a
deeper level, i.e. in the allocation logic of the program.

Moreover a diagram which is in an isomorphic 1:1 relationship with its
program, such as the block diagrams of GPSS, is likely to become cluttered
and thereby lose its immediacy. The main benefit of a diagram is that it

shows off the basic structures of a system in a readily assimilable fashion, and thus should not necessarily be complete in every detail: a diagram should be homomorphic rather than isomorphic to the program.

Besides the use of homomorphic diagrams as a readily understood description of the outline form of a model, the homomorph plays an important part in the engagement strategy, in the control of activations and data collection, as will be investigated in Chapter 8.

## 6.2 STRATEGY CHOICE
### 6.2.1 Strategy modularisation

A simulation program should be able to present the description of a dynamic system in a form which is both assimilable by the computer and comprehensible to the human writer and to whoever might want to read it. Comprehensibility is a laudable aim in any programming context, and is usually approached through the medium of a high-level language, developed for general purpose or specific problem areas. The nature of a high-level language is to offer a mode of expression which is convenient to the program writer; to shield him from the routine and automatic tasks which are necessary at the machine level, enabling him to concentrate on the accurate portrayal of the system.

Conventionally, we regard a simulation as a computer program which, when executed, behaves in a similar way to the real system the modeller has in mind. More precisely, the modeller maps the real dynamic onto the control flow of the program. Thus the principal difficulty of discrete-event simulation is the serialisation of parallel entity flows, i.e. the transformation of real parallel flows into a sequential flow of computer control (Tocher, 1969). This is accomplished by a choice of one of three simulation strategies: event-scheduling (ES), activity-scanning (AS) and process-interaction (PI). Any simulation program we write, whether using a language specially designed for simulation, or a general-purpose language like Fortran, will be committed to a particular strategy, or in some cases to a combination of strategies.

One important structural aspect of language design is the extent to which the programmer is encouraged to construct his program in terms of independent modules, by which we mean keeping separate the parts of the program which refer to different things, or functions. Modularity is commonly practised in general programming: subroutines, bracketed structures, etc., are the bread and butter of modern programming practice, and it is expressed through the programming language syntax. For simulation, the choice of strategy determines the types of module we employ. Real systems can often have a complicated structure, with component parts interconnecting in an irregular way. We require a modularisation flexible enough to capture this irregularity, while maintaining some program discipline.

Since the diagrammatic forms of representation, such as Petri nets, are not sequential, and therefore independent of strategy, we may use diagrams as a standard by which to judge the efficacy of each strategy. Let us consider the basic single-server queue, and how it is modularised by each strategy (Fig. 6.7).

The names of the ES-strategy modules could apply to the transitions of the net. The invocation of an event module corresponds to the firing of a transition and the subsequent removal of tokens from input places and their distribution among the output places. If the new arrangement of tokens enables further output firings, then this must be tested for explicitly from within the activated event module.

Inevitably, such propagation of testing leads to a poorly structured program since each module must concern itself with the possibility of further events being activated. In this example both 'arrival' and 'leave' modules must concern themselves with the invocation of 'start serve'. This is probably the main reason for the decline in popularity of ES-strategy languages. However, it remains common by virtue of the fact that simulationists using general-purpose languages like Fortran tend to adopt this strategy by default. But one shudders to think of how many ES simulations must have been written with crucial tests overlooked, and event occurrences omitted.

The AS strategy provides slightly more guidance for the simulationist in that it recognises the distinction between events which are solely dependent on the passage of time (B-activities) and those dependent on some aspect of the state of the model (C-activities). Alternatively one may describe the distinction as between direct and consequential actions. B-activities consist of procedures invoked directly from the event-scheduler, whereas the C-activities, being conditional, are constructed as modules containing a test-head followed by the actions to be taken if the test-head evaluates to **true**. This format allows the model to be specified in a more self-contained fashion without unnecessary duplication of tests. Every time a B-activity is invoked, a scan is made through all the C-activity test-heads, executing the action body whenever allowed, offering a crude but fairly effective way of making sure that *all* possible conditions are checked.

The PI strategy seeks to modularise at a higher level, that of the whole entity process rather than individual transitions. The process module comprises the total string of places and transitions traversed by an entity. A process thus indicates for each entity-type the sequence of states through which the entity will pass, and the other entities with which it may interact. Moreover, the possibilities of further conditional activations are, for the most part, catered for by following the sequence of actions within the process. Thus while the ES and AS strategies concern themselves only with tokens in places, the PI strategy also includes the sequential information given by colour-coded arcs. One long-term benefit of simulating with process modules is that, having more self-consistency, they are more convenient to store in a library (Birtwistle *et al.*, 1985).

### 6.2.2   Information content of a modularisation

In Fig. 6.7, the PI description is given in full, making use of the high level of constructs that the strategy allows. Notice that the ACQ and REL pair

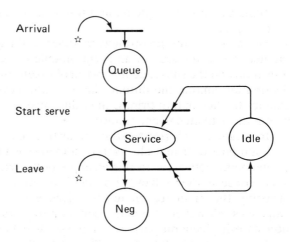

Modularisation by ES, AS and PI strategies:

ES: event modules: *arrival, start_serve, leave*

AS: B-activities: *arrival, leave,*
    C-activities: *start_serve*

PI: the customer process:

|          | GEN | *customer*; |
|----------|-----|-------------|
| *queue*: | ACQ | *server*;   |
| *service*: | TIME | *service*; |
|          | REL | *server*;   |
| *neg*:   | NEG | *customer*  |

Fig. 6.7 — Augmented Petri net for customer–server interaction with places and transitions named.

effectively enable the whole server cycle to be defaulted. This is because the server can be regarded as a typical resource-type entity, with a simple busy–idle cycle of activities. Needless to say, the ability to imply a whole process by a couple of high-level statements would be quite beyond the capabilities of the ES or AS strategies.

Most implementations of PI strategy are rather more sophisticated than those of ES or AS strategies, involving coroutine linkages between modules and the executive. For this reason alone it is too specialised to be implemented in most general-purpose languages, a fact which has largely impeded its acceptance in a micro-computer environment. The executive is governed by an FES as usual, but instead of a procedure invocation, a coroutine resumption is caused. In the majority of cases, there is no need to perform a scan of possible subsequent events once the particular entity concerned in the resumption is identified (Birtwistle, 1979). However, as we have seen in Chapter 4 there are still some pitfalls for PI, for example the unwitting inclusion of deadlock.

In respect of allowing a higher level of language and of modularisation, the PI strategy possesses distinct advantages over the other two. In addition, programs using PI are concise and, with the right orientation, easily understood. Ease of undertanding is of course a very subjective means of estimating a language, but it must be agreed that putting oneself in the position of a flowing entity and imagining the system evolve as it would appear to that entity does have distinct advantages over attempting to consider events in some arbitrary order. For instance, it is less likely that events would get omitted from the total description.

But it would be a mistake if the PI strategy regarded all interactions as typified by an entity–resource interactions. In a lift system, for instance, the lifts provide a service for the passenger, but the interaction of the two components is far more complex than between customer and server, with many passengers being served simultaneously, and with different requirements, all of which affect the service time of other passengers. So we need a more general means of describing interaction, for which the common assumption of entity–resource is a special case.

Without employing the defaults of ACQ and REL, the PI modularisation consists of two process descriptions: that of the customer and that of the server. The full program with no defaults is composed in terms of primitive statements from which we will later develop a simulation semantic model based upon engagements, the so-called engagement (ENG) strategy.

PI:  a customer process

|  |  |
|---|---|
|  | GEN *customer*; |
| *queue*: | WAIT UNTIL *server* IN idle BY FIFO; |
|  | COOPT *server* FROM *idle*; |
| *service*: | WAIT TIME *service*; |
|  | FORK *server* INTO *idle*; |
| *neg_customer*: | WAIT NEG *customer* |

and a server process

|  |  |
|---|---|
|  | FOR EACH *server* |
|  | DO CYCLE |
| *idle*: | WAIT JOIN *customer* IN *queue* |
|  | OD |

Despite its extensive use and promotion by proponents of the AS strategy, especially in the UK, the information in an activity-cycle diagram is more completely contained in a PI-strategy program: the only approach to take fully into account the triple of event, state and entity individuality. An additional benefit for the PI strategy should also be mentioned. Without being obscure, the PI module is very concise. If the ES and AS modules are represented in the form of procedures, which is what normally happens, then they will require extra encapsulation, adding to the textual length.

A comparison of strategies based on objective criterion, such as the preserving of properties from a Petri net, would indicate that the PI strategy is superior. Of course, this is not the only means of comparison possible. There are many other factors which can be taken into consideration, especially subjective factors, which can assume particular importance in specific cases (Hills, 1973; Virjo, 1972). Similar conclusions have been made or suggested elsewhere, but based on specific language-oriented criteria (Hooper and Reilly, 1982; Birtwistle *et al.*, 1985). So, as far as gathering together the information presented in the Petri-net model of a parallel dynamic system, we may rank the approaches

$$ES < AS < PI$$

in order of completeness.


## 6.3   PROGRAMMATIC REPRESENTATION
### 6.3.1   Packaging the executive

Before discussing how to write simulations by means of specialised simulation programming languages, we should first consider whether or not we really need a simulation language at all, especially given the sophistication of available general-purpose languages. Is there anything 'simulation-like' which hampers simulations being written in a general language?

If we ask the same question on behalf of another specialised area of programing, such as Artificial Intelligence (Campbell, 1984), we find that there are some techniques which may at one time have seemed peculiar to the specialism, but have later been absorbed into more general languages. The use of linear lists to represent sets, and the associated requirement for dynamic memory management, while originating partly in an AI context (LISP), have now become a component of many general-purpose languages, e.g. Pascal and Algol 68. Early progress on the set-handling facilities of SIMSCRIPT also demanded dynamic data management, leading to the buddy-system of garbage collection (Knuth, 1973b).

Another AI language, Prolog, is concerned with inference through a database of clauses. It uses pattern matching with backtracking in order to try to find a clause which 'fits' the current context. But the pattern-matching problem is not specific to AI; in fact the technique is used by the string-processing language SNOBOL (Griswold, 1975). So, by comparison with modern languages, perhaps there is nothing essentially 'intelligent' about an AI language.

Can a similar criticism be levelled at simulation languages? There are many ways in which simulation has stretched the capabilities of conventional languages, and acted as a stimulant to produce new algorithms. What is it that makes a simulation language so special? Firstly, whatever strategy we follow, we must have an executive program to drive the simulation along. At least, i.e. in the case of the ES strategy, this would consist of an FES with routines for the insertion of new future-event notices and the retrieval of the next event. Otherwise, we would be requiring the simulationist to provide this for himself, to re-invent the wheel, which might turn out to be square!

Convenience alone demands that the executive program should be part of the simulation language, although it would be out of place in a general context.

Secondly, the relationship between general and specific parts of a program is different for simulation than for more conventional, procedural languages. Conventionally, a calculation program is written in a general sense, in terms of variables whose values will be actualised at run-time, through a *read* statement. In this way a program resembles a mathematical formula like $f(x)$, with $x$ the unknown variable. The program thus describes the active procedure, $f$, which defines the way in which the eventual $x$ will be operated upon, while the data consists of passive $x$ instances. The function $f(x)$ produces a value, whatever $x$ is. This approach to modularising programs enables libraries of functions like $f$ to be built up and invoked when required.

But for simulation, we have already seen in Chapter 2 how the mathematisation of system concepts is too limited for practical use, in that simulation requires more than mere substitution in a prescribed formula; so we need the structures of the real system to be mirrored in a simulation language. In a simulation program, the generalisable aspect is the executive, applicable to the total set of discrete-event simulations. But the dynamic nature of simulation requires that the model instances are not passive; thus we must define a model capable of change, i.e. a dynamic model. Remember that it is the dynamic nature of simulation models which marks them off from database and calculational models. So now our 'data', by which we aim to instantiate the specific model of interest, consist not of mere passive values, but program modules of one kind or another which require activation. In addition, we may well require to specify data structures and subprogram definitions, etc., which are relevant to the particular model. These requirements cause a special difficulty which we will return to later.

Simulation programming languages should enable programs to communicate between programmer and machine on the one hand, and between different programmers, on the other. Ideally, a language should be close to natural or mathematical language (Aho and Ullman, 1977). We might add that for simulation, which is basically a mapping of a real dynamic onto the control structure of a program, we need a sufficient variety of control effects to capture a rich selection of possible real-system phenomena.

Simulations may be written in general-purpose languages only if we are prepared to accept restrictions. Set-handling for queues and temporary-entity data structures are generally difficult to provide conveniently in Fortran, which means that the whole program must be written in one piece, without separation of general and specific parts. Nevertheless an interesting attempt to do just this is given by MacDougall (1975) in the SMPL package. Crookes *et al.* (1986) provide a three-phase AS-strategy package in Pascal, where the difficulties of including procedures to instantiate the model are clearly shown, in that the user himself is required to write the code to perform the C-phase.

The simplest way in which a general package may be constructed is to write an FES-handler which will call from a set of event descriptions, provided by the user, as dictated by a pointer in the next-event data structure, giving rise to the event-driven programming style. The user-supplied event-procedure is then activated at the appropriate model-time. This is the basis of the ES strategy of discrete simulation. Its simplicity makes its construction very easy, and when a simulation is to be written in a general-purpose language, it is usually this strategy that is taken. It is also the most obvious, naive approach; but, as we have already seen, it is not necessarily the best. But the strategy may possibly be advantageous when a largely non-discrete-event simulation is being undertaken.

There may be sound practical reasons for attempting simulation in a general-purpose language. It might save learning another language — maybe several special functions have already been written in the current language, and there is a reluctance to change. This is especially so in the case of many combined discrete-event and continuous simulations, especially as the continuous part is more likely to be written in Fortran.

To sum up, simulations require a different kind of relationship between general and specific parts; the specifics are invoked by the general, which is the opposite case to conventional, algorithmic programs, where general library routines are invoked with specific parameter values. The only general-purpose language with sufficient flexibility for simulation is SIMULA, which started out initially as a simulation language, but now presents itself as a language for writing packages. Other object-oriented languages inspired by SIMULA are capable of supporting simulation (e.g. Smalltalk-80, Goldberg and Robson, 1983).

### 6.3.2   Level of language

The advantages of high-level languages have been frequently emphasised; few nowadays would criticise language designs for incorporating these properties:

(1)  ease of understanding of the program;
(2)  naturalness;
(3)  portability;
(4)  efficiency.

How high the level of a programming language is lies in the structural support it gives to programs. A Pascal programmer can take advantage of the concepts of type and data structure offered by the language as well as the choice, iteration and subprogram control structures. Less structural, but equally important, are the checking routines, e.g. array index within range, and use of built-in routines such as *sqrt* for finding the square root. These utilities absolve the user from writing standard procedures, and thereby from making mistakes, and allow him to concentrate on the programming problem in hand.

The control structures offer a different kind of support from the provision of utility subprograms. They supply guidelines which mould and guide the programmer's approach to any given problem. For instance, the iterative control structure based on the **for** keyword uses an enumerative type on which the index of the loop is based. But if the programmer finds it more convenient to advance the iteration in terms of a real (i.e. non-enumerative) variable he must reconstruct the control structure for himself, explicitly supplying the various tests, incrementation and jumps. Often the very existence of such a control structure will tend to encourage the programmer to enter the real index values in an array indexed by an enumerative type, e.g. an integer. In this way the existence of supporting facilities make its presence felt in the approach of the programmer.

But high-level languages are not without certain drawbacks. The most commonly cited disadvantage is that they lead to machine code which is overly inefficient. However, this is countered by considering the improvement in overall efficiency when the programmer's time, consistency, robustness and portability of code are taken into consideration. Less often do we hear that high-level languages can actually prevent some programs from being written. Some recognition of this point can be seen in the increasing use of languages like C (Schildt, 1986) in which the typing features are lax, allowing integers to be treated as pointers and so on. Of course there is always more than one way to write a particular program, so it is always difficult to assert that a program definitely *cannot* be written in a particular language. But there are of course many dangers lurking for the unwary programmer who decides to go for greater freedom. He must exercise his greater power with more responsibility, to ensure that he is not transgressing the coherence and integrity of his program.

In simulation, a high-level language should make the task of programming easier through the provision of structured forms of data and control Perrott *et al.* (1980). The necessity of employing a strategy will inevitably commit the user more or less to a specific world-view. In general, the more support provided, the more firmly the modeller is swayed to a particular modelling approach. But despite these strictures, no one seriously argues against the benefits of high-level simulation languages. Simulation languages have been regarded as offering the highest level of support. Even the earliest versions of GPSS contained the very powerful command to acquire a unit resource (called a facility in GPSS) such that the simple block statement

SEIZE *server*

would (a) declare '*server*' as a unit resource, or *facility*, (b) when the statement is activated, test if the server is currently available, and if so, convert its status to 'busy', (c) if it is currently busy, enter the activating entity into a FIFO queue, and (d) reactivate it automatically when it reaches the front of the queue and the facility becomes available.

But while the SEIZE command is very useful, like ACQ, it cannot cater for all kinds of inter-entity requests. A lower-level form of queueing, allowing priority acquisition of resources, is offered through user chains and their access blocks LINK and UNLINK, although automatic reactivation is not provided. By providing this support we are tacitly assuming a particular type of entity–resource interaction, which may need to be broadened.

At the lowest level of programming, the only control structures available are the **goto** and the interrupt (Enslow, 1974). Thus all control structures are ultimately reducible to these. But the idea of constructing large complicated programs on such a base is, to say the least, daunting and fraught with error. For various reasons, some authors prefer to use low-level languages. The best example of the benefits and drawbacks of such an approach are the simulations of the Caltech elevator written in MIX by Knuth (1973a). While being very explicit, in terms of implementation of coroutines and in being neutral towards any existing simulation language (although not neutral towards simulation strategy), it must be admitted that this simulation is hard to follow. Presumably the intention of the author is to demonstrate the details of the implementation, particularly as they involve coroutine linkages. Inevitably, high-level simulation languages will obscure such details.

### 6.3.3  Basic requirements of a simulation language

As with any programming language, we expect a convenient syntax defined in a programming manual, together with a more-or-less well-defined semantics, explained through examples. The simulation language should also support modularisation, with attendant rules about the intervisibility of variables and names. An adequate compiler and loader with comprehensive error-reporting at compilation and run-time, to ensure a logically consistent model, is of course also a necessity.

For the simulation-specific facilities, the *executive interface* requires a user-called routine which inserts future events into the FES, while time advance, i.e. the invocation of simulation modules by the next-event features of the executive, is usually implicit in the language. Similarly we require the access of entities to and from a queue or ordered set, defined according to either FIFO or LIFO discipline, or some other priority scheme. Some facility to search for items stored in sets, incorporating general set-handling features, is frequently useful. Dynamic data management will benefit from some kind of heap storage system, as in Algol 68, together with facilities to perform garbage collection in a non-intrusive way.

To obtain *chance effects*, we need some facilities for random-number generation, of integer and real type, with several streams. To incorporate probabilistic features we would require a comprehensive set of distributions from which to sample, e.g. exponential, uniform, Weibull, Erlang, geometric and Gaussian, together with a means to sample from histograms supplied by the simulationist, either in the program text or as data.

*Data-collection* facilities and automatic monitoring of queues, including mean, standard deviation of queue-length — both with and without zero contents — would be useful, as would automatic collection of entity transit times, and the number of transactions logged past a point. Ideally, these facilities would be positioned by separable data-probes. The format of the generated report could be specified; some form of histogram output is usually suitable. All aspects of data collection would be better specified outside the main description of the model.

We have already seen the advantages of the language supporting a *resource assumption* and the associated built-in operations like ACQ and REL. This simplifies the definition of many resources considerably. For more complex cases, the ability to define a data structure for entities, both temporary and permanent, and appropriate operators for collaboration between entities, is necessary. Lastly the interaction between the simulation and its computing environment, the possibility of interfacing with other programs and i/o devices, including real-time and graphical facilities, would be convenient.

An often neglected area of simulation language is the ability to perform *initialisation* of the model. The initial condition of a model can be extremely important in a 'what-if' exploration of feasibilities. Unfortunately to specify the internal constituents of sets, queues, etc., is rather tedious and awkward. Most likely the simulation would be given a 'cold start', i.e. basing the initial conditions on the assumption that all queues are empty, all resources idle, and so on, and the simulation would be run until the simulationist judges the level of activity as fairly typical of an initial configuration. At this point, statistics' collection is started. But such an approach is not always satisfactory. There is clearly a need for some facility to store a current configuration for subsequent re-initialisation.

In the dark days of batch processing, the limitations of experimentation were severely curtailed, because the results of each experiment must be considered before the setting up of the next run. For really effective experimentation, an *interactive environment* is essential for probing the potentialities of a system. In a 'what-if' mode of investigation, the simulation will probably be subject to a continuing series of slight amendments, which should be able to be imposed without incurring a total recompilation of the whole model. Monitoring of specific entities, for both experimental and debugging purposes, should also be regarded as separable. A major property of the language to enable this type of *incremental compilation* is the separation of data-probes and run-time aspects from the model itself. For this mode of operation, an interactive computer environment is a real boon to the simulationist.

Further discussion and some more detailed suggestions for simulation languages are given in the description of the SIMIAN project in Chapter 8.

## 6.4   METHODOLOGY

### 6.4.1   Intermediate model

For general programming, a preliminary diagram, such as a flow-chart, is normally recognised to be a beneficial stage in the development of a program, especially when it is composed of interrelated parts. In the case of simple programs, perhaps it is not so necessary. The wider availability of high-level programming software, like Pascal, has lessened the need for extensive diagrammatic rehearsal.

In simulation programming, the situation is more complicated, in that the flow-diagram describing a process must be understood as potentially re-enterable and executed in many concurrent instances. Viewed as a static object, the process might confusingly appear to interact with itself. In addition there will be interactions between the instances of different processes. Thus the visible sequence of the process is difficult to reconcile with the parallel reality. Petri nets, and other flow-diagrams specifically adapted for parallel system description, circumvent this problem by the use of tokens or counters to represent individual instances, and their position in the execution of their respective instance.

For a simulation programmer, the idea of modelling something as complex as, say, a container terminal, with only a language manual for guidance, would be quite unthinkable. For the simulation of designed systems, the argument for some explicit methodological process is greater. It was precisely as a methodological aid that Tocher's (1964) activity-cycle diagrams were invented.

Given the inherent advantages of a diagrammatic form of system representation, it is not surprising that many attempts have been made to generate simulation programs using diagrams as a source, attempting in the process to short-circuit the programming phase, which many people find difficult. In addition this procedure would absolve the modeller from any direct concern with the effects of strategy. Discussions on simulation methodology have therefore focused around the form of a model, which is *intermediate*, between the real system and the simulation program. As an intermediate model, it can be discussed and criticised by people who are not familiar with the syntax and facilities of a simulation language. By eschewing detail, different people may be able to perceive the overall structure of the simulation without it seeming to them to be over-complicated. By encouraging criticism from a wide range of expertise, more confidence may be gained that the simulated system will behave in correspondence with the projected real system.

### 6.4.2   Packaged methodologies

Following Hutchinson (1975), we describe the application of activity cycles to the methodology of simulation. After identifying the entity-types of interest, their behaviour patterns are mapped into sequences consisting of alternating states of activity and idleness. The idle states are thought of as queues, and the activities as productive states. Each entity-type thus has an activity cycle. When the interactions between entities are considered, some of the cycles will interact for certain activities. As opposed to the process-oriented diagram, such as that promoted by GPSS, it is not necessary to show the production flow, although in practice this will act as an orienting thread linking the cycles. Such a diagram enables C-activities to be identified as requiring cooperation between machines, as well as the entity-type they require; also the bound activities become clear. The complete diagram becomes a convenient basis for all with an interest to discuss the veracity of the model and its level of detail.

Once the model has been agreed upon, three additional types of information are required: activity durations, queueing disciplines and the starting conditions. Prior to presenting the model to the computer, the diagram can be used as a basis for simulation of the system by hand. The diagram in this case functions as a games' board with counters to represent entities. The game is played according to the following rules: for each activity in turn, decide if it can start, i.e. see if the input queues to the activity contain counters. If an activity can start, then the required number of counters is removed from the input queues and set down on the activity boxes. Then using the activity-duration information, calculate the end of activity event, and place this future time in the event set.

When all activities have been considered, the simulation can move on to the next event, which, of course, is the earliest in the event set. Counters are then moved to denote the effects of this event, and the activity scan is considered again. The cycle of operations is repeated until the finish time of the simulation has been exceeded, or everyone concerned is convinced that the simulation is behaving as the real system would. The rules of the game are nothing but a re-statement of Tocher's three-phase AS executive.

Two packages, HOCUS (Poole and Szymankiewicz, 1977) and ECSL-CAPS (Clementson, 1977), suggest the methodology should start out from an activity-cycle diagram. The model can then be hand-checked. At this point — the input of the model into the computer — they diverge: HOCUS prefers a structured input of data made up on standard forms, whereas CAPS allows a more flexible dialogue-form of input. Both approaches end up generating a Fortran program which is eventually compiled and run. A similar program generator, AUTOSIM (Warren *et al.*, 1985), has also been developed.

Mathewson's (1985) DRAFT is also program generator initiated by questionnaire, with provision to produce output in one of a set of simulation languages: SIMON, GASP II, SIMULA and SIMSCRIPT, incorporating any of the three strategies: ES, AS or PI. The generator can also be used to assist in the production of animated simulations, based on a Fortran implementation of the SIMON simulation language. Graphics is regarded as a natural enhancement to established program-generator software, having the potential to make simulations clearer to a wider circle of people. Another program generator MISDES (Davies, 1979) offers a choice of styles of questionnaire, in terms of event modules, an activity-cycle diagram or B- and C-activities.

Along with assistance in deriving or engineering a simulation program, intermediate models also function as documentation for the simulation. Note that for a simulation, documentation standards must be set so that all interested parties, from clients and top management, to systems analysts and applications programmers, can know the extent and intended validity of the simulation. Program documentation, on the other hand, can be briefer and more technical, the writer being able to presume a higher level of familiarity with programming language in the reader.

Going a step beyond the program-generator approach, Heidorn (1974) has suggested a package where the simulation is defined and refined in natural English, and a GPSS program is written in response. While posing as an exercise in natural-language understanding, it is clear that the scope of 'understanding' is severely restricted and that the programs produced are quite limited.

Each package has its individual pros and cons, but one general criticism can be levelled at the method of simulation by generating the program from diagrams. It seems that only the easy, simple models can really be adequately and completely modelled in diagrammatic terms. For really useful simulation, hacking away at a program is ultimately unavoidable. While most authors of packages honestly admit this, they counter with the assertion that at least their technique will bring about an error-free program that is 90% complete. The problem here is that it is the simplest 90% that has been done; to complete the job, the programmer must wade through the generated text making suitable amendments where he thinks necessary. This procedure does not inspire confidence in the result.

### 6.4.3   Methodological cycle

Zeigler (1976) was one of the first to propose a total scheme of computer modelling, from initial conception to final results. The procedure has five elements:

- real system,
- experimental frame,
- base model
- lumped model,
- computer.

The real system is the object of study, and is the source of observable data. Variables in the system may have inputs and outputs, corresponding to our idea of cause and effect, although the input–output behaviour is all we really know. The experimental frame serves to identify the limited set of circumstances under which the real system is to be experimented with. It is associated with a subset of the total input–output behaviour, which allows the question of model *validity* to be assessed. The base model is an assumed, complete, complex, and never fully known hypothetical explanation of the system's behaviour, whereas the lumped model is an approximation to it made by the experimenter lumping together components and simplifying the interactions. Finally the computer is regarded simply as a device to generate input–output pairs of the lumped model.

While bringing out certain important concepts, the separation of the experimental frame from the model, for instance, it must be said that the scheme is rather too theoretical. Perhaps the net has been cast too wide, to include all kinds of scientific investigation, as well as designed systems. The use of words like 'observable' and 'knowable' perhaps suggests a positivistic, physical preoccupation. A practical methology, based on a decomposition in terms of these functional elements, is given in Ören and Zeigler (1979).

Nance (1979) describes the so-called conical methodology, embodying a top-down definition of a simulation model, coupled with a bottom-up specification of the model. Definition starts with the objectives of the study, and the assumptions which relate to the objectives and a set of model attributes. The model itself is described as a hierarchy of submodels, with respect to a set of attributes, proceeding into deeper levels until no further subdivision is felt necessary, i.e. at the object level. The model specification starts at the object level, through association of objects to form submodels at higher levels in terms of assignments to the model attributes. In the use of a hierarchy, Nance hopes to enforce some kind of verification and validation on the model while it is being developed. Implementation of the conical methodology is through a Simulation Model Specification and Documentation Language (SMSDL), whose semantics are in terms of time and state relationships. Another simulation software engineering project, proposed upon similar lines, is the DELTA project based on Simula (Holbaek-Hanssen *et al.*, 1977).

The best description of simulation methodology is given by Overstreet and Nance (1985). Their concept of a cyclical repeat of definition, simulation

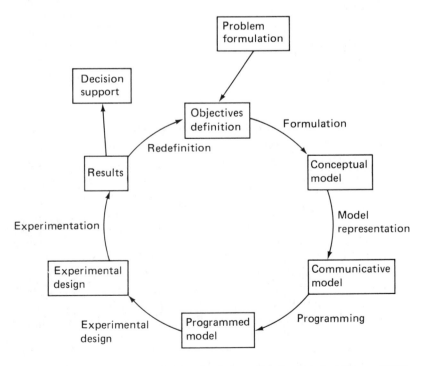

Fig. 6.8 — Methodological cycle for simulation, after Overstreet and Nance (1985).

and experimentation (see Fig. 6.8) neatly captures the idea of the modeller's typical 'what-if' attitude towards the system. First the problem is defined and the objectives decided, giving rise to a conceptual model, existing primarily in the imagination. To make this more concrete, a communicative model is drawn up which is of a non-technical nature and easy to understand by everyone involved. Typically, this would take the form of an intermediate activity-cycle diagram. Once this has been circulated and revised, a simulation is programmed, which is subjected to experimentation. The results from the experiment go towards the support of a decision, which could involve an interface with other computer systems. Further modelling may be necessary, in which case the cycle repeats.

The authors' aim in the paper is to put forward a model specification language called CS (Conditional Specification), a language in which a model *specification* is written. Once written, a specification can be subjected to analysis to see if there are any errors. The ultimate goal is to automate the whole procedure, based on the conical methodology, from specification to the experimental model. But as the 'conditions' of CS look exactly like C-activities, and the *set alarm* statement functions in an analogous way to our SCHEDULE statement defined in Chapter 4, it turns out that CS is little more than a simplified simulation language based on Tocher's three-phase AS strategy, from which it would be quite possible to compile a simulation program. Later work (Nance and Overstreet, 1986) suggests that the prime area of interest is in the diagnostic assistance that a CS permits.

There is a tendency to over-formalise methodological issues. All kinds of inconsistencies and opportunities arise when a system is being closely scrutinised, so a rigid adherence to a standard methodology might be counter-productive. Feyerabend (1978) is a refreshing antidote to formal methodology.

## 6.5  SOME SIMULATION LANGUAGES

### 6.5.1  Effect of strategy

The general trend of computer languages to offer a more extensive range of high-level concepts continues into the domain of simulation languages. Nowadays few would seriously argue against the advantages of high-level languages, although, with the great longevity of some simulation languages, low-level concepts still occupy too much consideration, as we shall see, and many so-called high-level languages still require a lot of detailed attention from the simulationist.

What we have said previously about the level of language confining the programmer to given forms of control structure, etc., applies more strongly in the case of simulation. As far as structural guidance is concerned there are the three established strategies: ES, AS and PI. We should consider therefore what kind of impediments are unwittingly taken on board by the modeller's accepting one of these strategies.

Paradoxically, the most supportive strategy (PI) also presents the strongest impediments to simulation. We have already seen how the lack of a means to express parallel multiple acquisition in PI can lead to an unwanted deadlock being brought into existence between contending entities. Also we have a problem where entity advancement is predicated on the existence of a particular model configuration.

In order to see and compare various features of simulation languages, let us take a brief look at their history. It seems that there were initially three independent centres of development, two in the USA and one in the UK. In the USA, the ES strategy was being promoted by SIMSCRIPT and GASP, and the transaction orientation by GPSS, whereas the British trend developed its own AS strategy, and developed a machine-oriented approach in the language GSP (Tocher and Owen, 1960), which led to further development of associated languages. At this time, simulation strategies were seen as either 'material based' or 'machine based', depending on whether the program took the viewpoint of the material being processed through the system, or the machines which performed the processing. In modern terminology these correspond to the PI strategy where attention is focused on the *flows* through the system, and the AS strategy in which the question of availability and need of resources are stressed.

The next phase of development was in response to innovations from outside the discipline. The concept of the coroutine, first used in the programming of a Cobol compiler (Conway, 1963), led to SOL (Knuth and McNeley, 1964a,b) and Simula (Dahl and Nygaard, 1966) which was also influenced by the trend towards language standardisation under the Algol umbrella.

The response to these developments was slow in the US. SIMSCRIPT became a language in its own right, instead of being just a preprocessor for Fortran, and adopted the PI strategy in version II.5 (Kiviat *et al.*, 1973) as an option. GASP stayed as a Fortran library and specialised its appeal as an ES strategy with an aptitude for combining with continuous simulations (Pritsker, 1979). It has since spawned some offspring, Q-GERT for network simulations and SLAM as an all-encompassing extension. GPSS developed along its own lines by expanding the number of types of statements, introducing some redundancy and overlapping (Gordon, 1978). GPSS and SIMSCRIPT will be discussed in later subsections.

### 6.5.2 AS-strategy languages

Tocher and Owen (1960) is the main point of reference for AS-strategy work in the early 1960s. They describe what was then regarded as 'automatic programming', which was simply a high-level provision for expressing a simulation. The language under development was GSP (the General Simulation Program) described in Tocher (1963, 1964). The approach is based on the concept of machines (resources) rather than material (entity) flow. The cycle of machine activity is composed of being *committed* to a particular duration of work, and then *returning* to be re-allocated, or perhaps becoming idle. The machine can be pre-allocated so that it may always repeat the same work, by being *engaged* to a particular activity. A major design aim of GSP being run-time efficiency, a program written in the language tends to resemble a cypher, with many single-letter identifiers and keywords. This obscurity, coupled with a recognition in industry of the potential benefits of simulation in planning, probably led to many variations on the GSP theme being developed in higher-level languages.

Buxton and Laski (1962) describe CSL, a Fortran-based language deriving primarily from the view that decision-making is based on the use of sets, the basic approach being therefore through the predicate calculus. Extensive set-handling facilities were offered, in a similar development to Markowitz's SIMSCRIPT, which was being developed around the same time. It was portrayed as a general language, with first application in the field of simulation, with facilities based on those of Tocher and Owen (1960). Buxton (1966) describes a second version, more specifically about simulation. The two-phase time-cell variant of the AS strategy is described. The author refers to the possibility of recycling the set of activities after an activity has been entered, and of the necessity of a careful choice over the order in which activities are specified.

Clementson (1966) produced an extension of CSL, called ECSL. Since compilation into Fortran was found to be too restrictive, an ECSL compiler was written to compile directly into machine code. ECSL adds two sections to the original CSL, for finalisation and for data. Pidd (1984) gives a comprehensive example of an ECSL simulation, and a description of using a CAPS dialogue to generate it.

Williams (1963) describes an Algol-based language, ESP. The author elucidates well the central problems of discrete-event simulation. The main problem is the control of the sequence of inter-dependent actions. The division between time- and state-dependent events (delayed and conditional actions) is also highlighted. ESP takes the form of a set of Algol procedures, based on the three-phase AS strategy. The simulationist communicates the names of his activities through an Algol **switch** statement

**switch** *act* := *maybe, first, second, third*;

containing an array of labels, denoting a conditional action, *maybe*, and three delayed actions: *first*, *second* and *third*, to enable the executive to invoke supplied labelled actions. The sequencing of the program is performed by a statement

**goto** *act* [*next*];

where *next* is an **integer procedure** which selects the next action required. Each supplied action should be labelled with an element of the *act* switch list, and terminate with this **goto** statement to allow the executive to control the sequence. A crude system of interacting semi-coroutines is thus defined in Algol. By comparison with simultaneous work based on Fortran, the central problem of sequencing parallel actions has been clearly focused. In connection with ESP, the heap structure for the FES was being investigated (Williams, 1964). Another Algol-based simulation language, Simon, described by Hills (1967), is also derived from the three-phase AS strategy of Tocher (1963).

Alongside the development of AS-strategy languages, developments in the PI strategy were taking place in the form of SOL (Knuth and McNeley, 1964a,b) and Simula (Dahl and Nygaard, 1966), and the emergence of a new standard version of Algol, Algol 68. This state of affairs is reflected in the IFIP symposium on simulation programming languages (Buxton, 1968). Many topics, including simulation diagrams, garbage collection, language standardisation, implementing the 'wait-until' statement, and quasi-parallel programming, come up for discussion, some of which are hotly debated today.

Shearn (1975) attempts to produce a simulation package ALGOLSIM using Algol 68 as a basis. A primitive version of the three-phase AS strategy is developed. Extensive use is made of the Algol 68 facilities for union modes, operator declaration, record handling and a sentential sequence of operators to perform set-handling. The packaging of the executive is, however, rather clumsy, every simulation project requiring its own library made up from an integration of standard modules and declarations with those pertaining to the particular simulation. This lack of ability to abstract the data types needed in the simulation also hampers the use of Algol 68's extensive dynamic storage management system.

Significantly, some of the areas of Algol 68 which have been considered for inclusion into the language, but rejected, are those which would affect its use as a simulation executive (van der Meulen, 1977). These include the ability to hide certain variables from the user of a library, the use of mode alternatives (modals) to enable, say, general queues to be set up for items of arbitrary data structure (Lindsey, 1974b); and partial parameterisation whereby access to a queue could be restricted to the declared discipline (Lindsay, 1974a). The difficulties with reconciling well-structured programs with a user-friendly environment for experimentation have meant that implementations of AS-strategy executives have tended to be in the form of program generators. With the popularisation of object-oriented environments, supporting abstract data types, incremental compilation and sophisticated graphics features, this tendency may decrease, especially when its flexibility and capability to handle events with configurational dependency are fully appreciated.

It is noticeable that most work on the AS strategy has been carried out in the UK; in fact (Tocher, 1979) it has largely been ignored in the USA, except for the work of Overstreet and Nance (1985), and Hooper (1986).

Before leaving the topic of AS strategy, we should mention the similarity between a collection of C-activities and a set of *guarded commands* (Dijkstra, 1975). Although they share the same format of test-head followed by action body, the similarity is superficial only: *all* C-activity bodies with test-heads evaluating to **true** should be executed, whereas for a set of **true** guards, whichever body is executed is not determined.

### 6.5.3  SIMSCRIPT

SIMSCRIPT started as a high-level simulation language based on the ES strategy. The simulationist defines his model in terms of entities and event

routines. Individual members of an entity class may have specific values
assigned to their parameters. A group of individual entities with various
properties in common may form a set. This means that the complete
configuration of a model at any time is given by the current list of individual
entities, their attributes and set memberships. The events, which bring
about the instantaneous state-changes, manifest themselves in the amend-
ment of either attribute values or set membership, or the generation or
termination of individual entities.

In the earliest version (Markowitz *et al.*, 1963), SIMSCRIPT was originally conceived as an extension to Fortran, but is now a language in its own
right. Its style is governed by an attempt to render the language as English-
like as possible, giving rise to a rich sentential structure of statements and
allowing many ways of describing the same effect. Entities are divided into
permanent and temporary categories, the space occupied by temporary
entities being cycled back for re-use. The modern version, SIMSCRIPT
II.5, has its own compiler and incorporates a process-interaction strategy. In
fact ES and PI strategies can be mixed freely in the latest version (Russell,
1983).

Every SIMSCRIPT program is composed of a sequence of segments: the
preamble, then the main part, followed by the processes and associated
routines. In our example (Fig. 6.9), the preamble declares two processes
corresponding to the definition of customer and customer generator. The
busy–idle cycle of the server enables us to model it as a resource. In the main
segment the server resource is initialised by the CREATE statement, in
which the number of kinds of server is set to one. The numerical attribute U
for the SERVER(1) resource is then used to set the number of server units to
two, specifying that there are two identical server resources, both attending
the same queue. The ACTIVATE statement creates a generator-process
instance and activates it at the current simulation time (i.e. immediately).
START SIMULATION indicates that all initialisation is complete and that
control should pass to the FES-handling routines.

The generator process creates 500 customer process instances at an inter-
arrival rate determined by sampling from an exponential distribution with
mean 4 minutes. Random-number stream 1 is used for this sampling, as
specified in the second parameter. The WAIT statement controls the inter-
arrival period between each customer. The customer process indicates that
each customer will require a server, which, if unavailable, will cause the
customer to wait in a FIFO queue until the customer's turn to be served
comes round.

Once a server is acquired, the service period is determined by another
random sample from a uniform distribution between 2 and 6 minutes. Notice
that here we have chosen to use the keyword WORK rather than WAIT
which was used for the inter-arrival periods. In SIMSCRIPT these words are
interchangeable, which allows the programmer to use whichever seems
more readable in the context. After the service activity, the server is
relinquished and either made to go into an idle state, or made available to
the first customer in the queue. In the latter case the waiting customer is
automatically reactivated and enters the service activity. After releasing the

```
PREAMBLE
    PROCESSES INCLUDE GENERATOR AND
        CUSTOMER
    RESOURCES INCLUDE SERVER
END

MAIN
    CREATE EVERY SERVER(1)
    LET U.SERVER(1) = 2
    ACTIVATE A GENERATOR NOW
    START SIMULATION
END

PROCESS GENERATOR
    FOR I = 1 TO 500,
    DO
        ACTIVATE A CUSTOMER NOW
        WAIT EXPONENTIAL.F(4.0, 1) MINUTES
    LOOP
END

PROCESS CUSTOMER
    REQUEST 1 SERVER(1)
    WORK UNIFORM.F(2.0, 6.0, 2) MINUTES
    RELINQUISH 1 SERVER(1)
END
```

Fig. 6.9 — Simple customer–two-server simulation in SIMSCRIPT II.5.

server, the customer is removed from the simulation and the memory space made available for reallocation.

As written, the simulation will produce no output because the programmer has not requested information on system behaviour to be collected. This may be done by an ACCUMULATE statement placed in the preamble, indicating which particular aspects are to be monitored, and a PRINT statement in the main segment, which defines the output format. Queue-performance statistics such as average, variance and maximum queue-length, and resource use are available as standard features.

Overall, a SIMSCRIPT simulation, especially when full use of the process concept is made, results in a fairly lucid and readable program, if somewhat long-winded. Comprehensibility is chiefly burdened by the intrusion of data-probes and output formats and the need for the reader to memorise the meaning of numerical attributes.

### 6.5.4  GPSS

GPSS is a high-level simulation language, and, as we have seen in the description of its SEIZE command, can embody a great deal of routine programming in a single statement. It is the main simulation software supported by the IBM corporation, and is thereby assured of widespread use. Documentation is extensive, and books and manuals describing its use abound, the best being written by its originator, Gordon (1975).

According to Gordon (1978), GPSS was initially acceptable as a programming language because engineers felt happy with diagrams. GPSS could be seen simply as a language for writing diagrams for the computer, every GPSS statement having a corresponding symbol, and every program its corresponding diagram. However, the GPSS diagram is not as useful as a development tool, because it is as difficult to draw as to write the program.

Strangely, despite being one of the first discrete-event languages, the word 'event' is hardly used, and, although being definitely in the PI strategy, prefers to call itself transaction-based. The transaction, shortened to Xact, is more or less what we call the entity, which flows through the system to produce dynamic effects. In GPSS all Xacts are GENERATEd and form the only flow of control available in the program.

The Xacts undergo interaction with resources by means of flowing into *blocks*, or program statements. A block remains inactive until an Xact attempts to enter it; some block types can prevent an Xact entering, in which case the Xact joins, by default, a FIFO queue. In this way resource-limitation can be expressed. Over 40 different blocks describe resource requirements, conditional branching; many kinds of queueing; data collection; report generation and attribute control.

Since all flow of control must be on behalf of an Xact, the experimental frame is completely interwoven into the model, as is the allocation logic. A resource is either a facility if there is only one of them, or a storage, of which there may be many; each type has its own blocks for acquisition and release. Xacts may have attributes, but reference to them often has to take into consideration their machine representation. Standard numerical attributes are available to obtain information on the past history and current status of the model. Searching and looping must be by Xact flow, frequently requiring the invention of 'fictitious' entities. The translation into machine code is achieved by line-by-line interpretation. While being slow, this method has the advantage that changes in the model can take place without too much disturbance.

The most immediate impression that a GPSS simulation gives is its extreme brevity, at least for simple models (Fig. 6.10). Moreover, by comparison with SIMSCRIPT the simulation gets straight on with the description of the customer process without the need of preliminary declarative segments. Typical GPSS processes are delimited by GENERATE and TERMINATE blocks, the inter-arrival information being given as a parameter to the GENERATE block. To denote the resource acquisition, the ENTER statement requests one item of storage (resource). Since it is this acquisition which would be the focus of interest in the model, we enclose the block between QUEUE and DEPART blocks to ensure that the statistics referring to the queue WAIT are collected. Once a server is acquired, the service activity proceeds for a duration determined by a random sample

```
        SIMULATE
*
        GENERATE     400,FN$EXP
        QUEUE        WAIT
        ENTER        SERVER
        DEPART       WAIT
        ADVANCE      400,200
        LEAVE        SERVER
        TERMINATE    1
*
        STORAGE      S$SERVER,2
        START        500
        END
```

Fig. 6.10 — Simple customer–two-server simulation in GPSS V.

from a uniform distribution in the range 400 ± 200 time units. Notice that
GPSS possesses no 'natural' time units, so the simulationist must choose an
appropriate scale. Here we use hundredths of a minute as the integer time
unit. The server resource is released by a LEAVE statement, and the
customer then leaves the system, and by so doing increments the termination count by 1.

Since the resource in this case consists of two servers, both working with
the same queue, the STORAGE form of resource is used. The synchronisation pair ENTER and LEAVE apply to this type of resource. If there were
only one server, then a facility resource could have been used, with
synchronisations SEIZE and RELEASE. Control of the simulation experiment is subsequent to the main model description, and prescribes a run of
500 customers through the TERMINATE statement. The use of the
QUEUE and DEPART data-probes gives us statistics on the WAIT queue
behaviour: maximum length, average length, total entries and average time
per transaction are obtained. These statistics are also calculated for the case
where zero-length queues are excluded from the reckoning.

The attribution of a 'high level' to a simulation language should be made
cautiously. GPSS, for example, while including high-level concepts such as
SEIZE is quite low level in, say, the specification of attributes. Any
assignment of attribute value must include some reference to their type
and/or their bit-size. All arithmetic computations in GPSS involve similar
awkwardness. These necessities are an unwarranted inconvenience. Other
low-level features persist in GPSS. The way in which attributes are refer-
enced, involving a specifically numbered sequence of commas, is poor
compared with the use of a field name.

GPSS is widely used and is implemented on many machines. Some
implementations are quite different from the GPSS V described here. We
should mention a Fortran implementation (Schmidt, 1978), and a micro-
computer version GPSS/H (Henriksen and Crain, 1982).

### 6.5.5  Simula

SIMULA started life as a preprocessor to Algol 60 with which to perform simulation programming (Dahl and Nygaard, 1966), then branched out as a package-writing language (Birtwistle *et al.*, 1973), and currently sets its lights among the new batch of languages embodying object-oriented concepts. SIMULA added the CLASS concept, coroutines, references, and record structures to the Algol 60 base, while omitting the troublesome **own** and **string** concepts.

The central idea of Simula is that of the *object*. An object is used to represent the behaviour of a major dynamic component of the system under study, corresponding to what we have called an entity. An object has to reflect the features of the actual component which are judged to be relevant to the purpose of the model — not only its static properties (entity-attributes), but also the actions it carries out as it moves through the system (sequence of activities). Normally we find that the many objects which occur in a dynamic system fall into a smaller number of sets of objects with similar behaviour patterns. A generic definition of an object is made through a CLASS declaration which contains both attribute and activity sequence definitions, each class of objects requiring its own CLASS declaration. Many objects of the same class can co-exist, cooperate and compete with one another.

Objects come into existence by a call of NEW followed by the class name, followed by an optional list of parameter values by which the attributes may be initialised. Then the activity sequence of the object is entered, until the sequence comes to an end, when the object is regarded as terminated. Besides having many objects of the same class existing simultaneously, many simulations demand objects which partially resemble one another. Two classes may perhaps share some, but not all, of their features. Simula provides a mechanism for defining subclasses of a class where specific differences between subclasses can be defined without repeating the features which they share in a common class declaration. Not only does this facility avoid duplication in the definition of classes, but it enables the same object to be identified either as a member of its own class or as a member of its superclass, the class by which its own is prefixed.

For example, when we consider road traffic, it is convenient to be able to define objects to represent cars, lorries (trucks) buses, motorbikes, etc. All types of vehicles will have licence plates, year of manufacture, a driver, and so on, but we may want to distinguish some qualities too — the unladen weight of a lorry or the carrying capacity of a bus may be significant attributes. In Simula we can define a class *vehicle* which contains all the attributes that vehicles have in common, and then define individual classes: *car*, *bus*, and so on with their particular properties of interest, as subclasses of *vehicle*. The class name *vehicle* then prefixes each subclass declaration, and the subclasses inherit the common properties of the superclass, which are merged with their own properties.

Thus we have in Simula a capability of abstracting a data type: we can think of, and program, an individual car as either an instance of a *car* class or a *vehicle* class, whichever is convenient. To simulate a queue of traffic waiting for a ferry to arrive at a pier then a queue of mixed subclass type, but having a common prefixed class, may be set up, containing cars, buses, lorries or motorbikes, in any combination, as long as they are all vehicles.

As far as simulation is concerned, the Simula concept of object may be equated with what we have called an entity. But in fact the object is much more general and can be applied to the flow of control within the program itself. Furthermore a CLASS inclusion hierarchy may be defined in such a way that built-in classes may include user-written classes which can make use of standard procedures. Thus the *simulation* class is a class enabling users to compose simulations making use of procedures like activate and passivate, etc. By calling the *hold* routine in their activity sequence, they allow the system FES scheduler to resume them as coroutines.

In Simula, the process as coroutine is typified in terms of a CLASS of which many instances, corresponding to individual entities, may be in existence simultaneously, enabling, in the case of the *simulation* class, a simulation of parallel processes. In place of the resume statement, we have three sequencing procedures: *hold*, *passivate* and *activate* (Fig. 6.11). A good introduction to objects and class hierarchy in Simula is given in Ichbiah and Morse (1972).

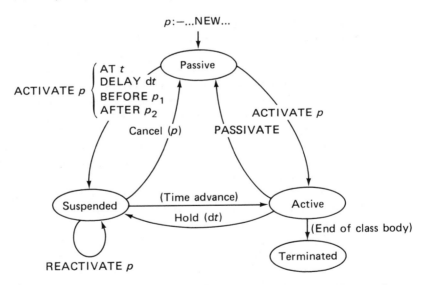

Fig. 6.11 — The states of a Simula process, and the transitions between them. *p* is a reference to a process instance, as are $p_1$ and $p_2$. *t* is an absolute time, and dt is a time duration. From Lamprecht (1983).

The *hold* statement corresponds to the GPSS ADVANCE statement and expresses the fact that the entity is currently pre-occupied and will remain so, without need of further attention, for a period of time specified in the parameter value. Thus the hold statement causes a resume of another process instance, as yet unknown. The controlling superstructure decides which one to enter next; of course, it is the one associated with the earliest member of the FES. In addition, an event-notice referring to the process instance in which the hold occurred is inserted into the FES, with an appropriate event-time, i.e. to denote the end of the holding period. When this time occurs the instance is automatically restarted at its reactivation point, immediately after the hold statement.

The *passivate* statement also halts execution of the current instance, but does not re-schedule it. The next process instance from the FES is resumed, while the passivated one waits explicit reactivation in an *activate* statement. Consequently, Simula simulations often require much activation AT NOW (Birtwistle *et al.*, 1973).

Let us consider the two-server supermarket queue written in Simula. The whole model (Fig. 6.12) is defined as a subclass of the built-in Simula class called *simulation*, with three processes: *customer*, *server* and *custgen*, a report procedure, and an initialisation body. The supermarket is parameterised by $n$, the number of servers. New customers enter the queue, whose maximum length is monitored, and activate a server in the idle state, if there is one. If not, the customer waits in the queue in FIFO order. Once the customer has succeeded in activating a server, the customer is passivated and its future is transferred to the activities of the server, only to resume control to calculate its elapsed time and to add one to the total number of customers. At the end of the process, the customer is terminated.

The irony is that while SIMULA is undoubtedly a high-level language, as far as simulation is concerned, it does not provide as much support in some areas as either SIMSCRIPT or GPSS. However, the basic components supplied by SIMULA enable the simulationist to write packages which are more closely integrated with his own problem area. For instance, Birtwistle *et al.* (1973) provide an example package for simulating a wide range of situations involving a service station which handles a stream of customers. In the language DEMOS (Birtwistle, 1979), the package idea is taken further to embody a SIMULA-based language with automatic reactivation for queueing entities.

Consider the same model written with a DEMOS prefix (Fig. 6.13). Here the resource-limitation is expressed through the ACQUIRE–RELEASE pair of synchronisation statements, much as in the case of GPSS. Another pair of synchronisations, TAKE and GIVE operating on the BIN SHUT-DOWN, limits the extent of the simulation to 500 customers. Notice that none of the basic SIMULA synchronisations is used; in fact, DEMOS is not a subclass of the SIMULATION class, but a higher-level alternative to it. Altogether five DEMOS synchronisation pairs are available, completely remodelled on the primitive synchronisations GET and PUT, which form a basic DEMOS reactivation model. We will consider this language more closely in our discussion of simulation complexity in Chapter 7.

### 6.5.6 Comparison of languages

We have already discussed the importance of strategy when comparing simulation languages. Many people feel at home with one kind of strategy

```
simulation CLASS supermarket (n); INTEGER n;
BEGIN process CLASS customer;
    BEGIN REAL entrytime, elapsedtime;
        INTEGER qlength;
        entrytime := time;
        into (queue);
        qlength := queue.cardinal;
        IF maxlength < qlength
            THEN maxlength := qlength;
        IF idle.empty
            THEN ACTIVATE idle.first;
        passivate;
        elapsedtime := time−entrytime;
        ncust := ncust+1;
    END***customer***;

    process CLASS server;
    BEGIN REF (customer)served;
        WHILE TRUE DO
        BEGIN out;
            WHILE queue.empty DO
            BEGIN served :− queue. first;
                served.out;
                hold (uniform(2.0,6.0));
                ACTIVATE served;
            END;
            wait(idle);
        END;
    END***server***;

    process CLASS custgen;
    BEGIN WHILE ncust ≤ 499 DO
        BEGIN ACTIVATE NEW cust;
            hold(negexp(4.0,u));
        END;
    END***custgen***;

    PROCEDURE report;
    BEGIN
        . . .
    END***report***;

    REF(head)idle,queue;
    INTEGER ncust, maxlength, u, i;
    idle:− NEW head;
    queue :− NEW head;
    u := inint;
    FOR i :=1 STEP 1 UNTIL n DO
        NEW server.into(idle);
END***supermarket***
```

Fig. 6.12 — Simple customer–two-server simulation in Simula.

```
BEGIN EXTERNAL CLASS DEMOS;
DEMOS
    BEGIN REF(RES)SERVERS;
           REF(RDIST)NEXT;
           REF(BIN)SHUTDOWN;
           INTEGER NCUST;
    ENTITY CLASS CUSTOMER;
       BEGIN
           NEW CUSTOMER ("CUSTOMER").
               SCHEDULE (NEXT.SAMPLE);
           SERVERS.ACQUIRE(1);
           HOLD (UNIFORM (2.0,6.0));
           SERVERS.RELEASE(1);
           NCUST := NCUST+1;
           IF NCUST = 500 THEN
               SHUTDOWN.GIVE(1);
       END***CUSTOMER***;

       SERVERS     :- NEW RES("SERVERS",2);
       SHUTDOWN :- NEW BIN("SHUTDOWN",0);
       NEXT        :- NEW NEGEXP("NEXT",4.0);
       NCUST := 0;
       NEW CUSTOMER("CUSTOMER").
           SCHEDULE(0.0);
       SHUTDOWN .TAKE(1);
    END;
END;
```

Fig. 6.13 — Simple customer–two-server simulation in DEMOS.

and resist taking up an alternative. However, there does seem to be some indication that the PI strategy has definite advantages, as we have remarked in section 6.2. But there are other factors to consider in a language comparison.

The format in which languages present themselves to the user is another axis along which we may compare their relative merits. Pidd (1984) describes four ways in which designers have approached simulation-language presentation:

- as collections of subprograms;
- as languages with statement-descriptions;
- as languages based on flow-diagrams;
- as interactive program generators.

It is true that some simulation 'languages' are nothing more than a library of Fortran subprograms (e.g. GASP, SMPL, etc.), whereas others are 'true' simulation languages such as SIMULA or SIMSCRIPT. In the third category of flow-diagram languages, Pidd lumps together ECSL and GPSS, languages which cannot be said to have much in common, especially since, as Greenberg (1972) points out, the flow-diagrams of GPSS are hardly necessary. The final category is those languages which invite the modeller to an interactive dialogue for model description (CAPS, DRAFT). The categorisation does have some merit, however, in identifying an approximate direction of progression of languages, from subroutine libraries, through self-contained languages to systems which embrace the whole user environment.

One of the first papers to call attention to the development of simulation languages was Krasnow and Merikallio (1964), who compared SIMSCRIPT, CSL, GPSS, SIMPAC, and DYNAMO. In their discussion they considered such topics as the range of arithmetic operators and set-handling. The conclusions were fairly equivocal, and their vista of the future was that languages would become more flexible.

Tocher (1965) compares several leading languages of the time, GPSS, SIMSCRIPT, early SIMULA, CSL, MONTECODE (Kelley and Buxton, 1962), and SIMON, with his own creation GSP. Several dimensions were used in the comparison: machine or material orientation, method of time advance, naming of entities, searching, subprogram writing, sampling, statistics, collection, initialisation and monitoring. Teichroew and Lubin (1966) extend Tocher's comparison by presenting a rather more thorough comparison, including a detailed comparison of SIMSCRIPT, CLP, CSL, GASP, GPSS and SOL. Five large tables comparing aspects such as data structures, various programming features, compilation, diagnostics, etc., are presented. Both comparisons conclude that while there is no realistic choice available to the simulationist, since this is usually made for him by the machine available, some discussion of simulation-language features provides motivation for their comparison and improvement. A detailed discussion of simulation languages, covering the topics of objects, terminology, sequencing, and set-handling, together with a comparison of GPSS, SIMSCRIPT, CSL, SOL, and early SIMULA, is given in Dahl (1968).

A study of three languages and their use in the simulation of three typical models has been published by Virjo (1972). By this time it had presumably been realised that previous comparisons had been rather equivocal, and that in any case, all well-known simulation languages seemed to offer much the same facilities, packaged in different ways, so the important task was to discover whether there was any explicit advice to give to the potential user. The three languages are GPSS, SIMSCRIPT and SIMULA, the most commonly used languages at the time. The three models are a simple machine shop with queues, a telephone exchange without queues, and a more complex stochastic PERT network. The overall conclusion was in favour of SIMULA, with a wish for higher-level statistics-collection facilities, like those of SIMSCRIPT, but the author was skeptical about the language's future as it is not backed by a large organisation.

The idea of comparing languages by their run-time efficiency and memory requirements, as they perform the same simulation, is continued by Musielak and Stoessel (1979), who compare GPSS-Fortran, GPSS V and

GASP IV in the simulation of a computer system, a doctor's waiting room and a lift system. The authors conclude by preferring GPSS-Fortran and GASP IV over GPSS V because of the latter's limited application area. A comparison of GPSS and Simula on a UNIVAC machine (Niederreichholz and Stockheim, 1979) found GPSS better only for simple models; as complexity increased, the Simula version of the test models was superior.

Kindler (1981) develops an algebraic theory of simulation-language concepts, of discrete-event, continuous and combined types, to arrive at a conceptual framework upon which to discuss simulation in general, and to compare languages. His classification of languages is concerned with which elements are considered as being active, which elementary, and whether there are B- and C-phases in the executive. The theory is currently under development.

There are many ways in which one can compare simulation languages. One may consider their efficiencies, both at run-time and from the global system-development viewpoint; whether they can perform continuous modelling concurrently with a discrete system; or their reporting and debugging facilities. Sometimes more subjective appraisals, such as ease of conceptualisation, are valuable.

### 6.5.7   Conclusion

This superficial look at the variety of simulation languages should be enough to convince the reader that there are simply too many simulation languages. It seems that the inherent difficulties of simulation have led to many attempts to reconcile various ideals, such as user-friendliness, portability, efficiency, standardisation, and to be a commercial success, all at the same time. Facing the choice can be bewildering for the first-time user.

What can be done to improve the situation? Piecemeal improvements have been largely responsible for the current plethora. One way of proceeding is to identify more language components with which we can capture complex phenomena, which is the approach we take in the next chapter. The typical 'what-if' relationship that the simulationist has with the simulation is also problematic, as it demands close control over the model, with frequent alterations, and flexibility over what measurements to make. Some intelligent guidance with system properties, such as contention, would also be advantageous. We address these problems in Chapter 8.

# 7
# Simulation complexity

" 'All management problems are of one of the seven types.' This is the most arrant nonsense." — from Rivett (1980), page 43.

## 7.1  WHAT IS SIMULATION COMPLEXITY?

It is a common complaint among simulationists that the systems they are trying to represent are much more complicated than the examples given in textbooks. Many extensive compendia of simulation examples exist, e.g. Schriber (1974), but naturally they tend to show off features of the language on offer, rather than give much practical advice about tackling an awkward problem. In part, such a selection of problems is inevitable, the author of an introduction not wanting to cloud his points with extraneous detail, but it is also perhaps indicative of an uncertainty or unfamiliarity with handling complex problems. Nor do examples of complex simulation receive wide publication for similar reasons: no one wants to advertise their difficulties. In this chapter we make an attempt at characterising the types of complexity so that we can make some comparison between simulation languages at a higher level of intricacy.

The first question to answer is: 'What makes a system complex?' or 'Where do we draw the line between simple and complex simulations?'. We should first mention that we are not concerned here directly with algorithmic complexity, which is a way of considering run-time efficiency, but a complexity of representation.

A convenient starting point of simulation complexity, and the one adopted in the introduction to HOCUS (Poole and Szymankiewicz, 1977), is the distinction between entities which are individually distinguishable, and those which are identical members of a set. Simple simulations are thus solely concerned with gross features of flow such as throughput and resource use, for which the individuality of entities is not relevant. Entities are individualised when they possess attributes, the values of which can be used to describe more closely the qualities of each entity, and to define more precisely the nature of their interactions.

Without attributes, all entities of the same class are regarded as just so many repetitions. By acquiring attributes, an entity can have more of an individual character, and can be involved in a more specific interaction with its collaborating entities. The precise details of the interaction will generally be difficult to include in a diagram. Significantly, simulation-program generators, which claim to handle only relatively simple models, have only limited facilities for the representation of entity attributes (Davies, 1979).

There has been a tendency to understand complex systems in terms of a hierarchy (Simon, 1973). In our view, while recognising the existence of systems at various hierarchical levels is important, a hierarchical view should not deny interaction between different levels. When we consider the highest form of complexity, the interrupt, we shall see that strict hierarchical assumptions tend to impede the modelling capability. However, the development of Simula has led to the setting up of *class* hierarchies, in which subclasses can *extend*, rather than restrict the properties of the superclass.

Zeigler (1984a,b) considers complexity as an intuitive difficulty in understanding the structure and behaviour of the system under study, arising from presumed interactions of a system's components. He seeks a global view of a system, untrammelled by conventional distinctions between scientific disciplines, and considers three main complexity-reducing operations; system boundary specification, decomposition and aggregation, thus seeking to deal with complexity at the level of the conceptual model.

By starting out from simulation language, we are aiming at a bottom-up understanding of complexity, by extending our powers of description to embrace more complex control features. Our characterisation of complexity contains seven categories which form into two groups: linear and non-linear. Linear complexity is concerned mostly with the details of entity development in time, partaking in queues, etc., as a linear sequence of happenings. On the other hand, non-linear complexity considers the possibility of departures from a linear sequence, enabling entities to evolve on more than one line, splitting and merging and even interacting with different aspects of their own behaviour. An earlier characterisation of complexity is given in Evans (1986a).

Based on our complexity characterisation we can take another look at existing simulation languages and compare them on their capability to handle complexity. In taking this approach to language comparison we are concerned that the simulationist should not be forced to indulge in 'subterfuges', or devious use of language, in order to formulate the simulation. Of course it would be rash to conclude that once a language can grapple with these complexities that we have somehow reached the end of the road as far as language development goes. Further difficulties will be discussed in connection with the development of the SIMIAN language in Chapter 8.

## 7.2  LINEAR COMPLEXITY

Following the first paradigm of discrete-event simulation, our characterisation of complexity concentrates on the description of the transitions of the model occurring at discrete events, when entities may be dissolved from current interactions and are able to enter into further *engagements* with one another. Our concerns are then with *allocating* entities to one another to perform collaborative engagements, together with defining the extent of new engagements.

We prefer the term 'engagement' over the more common 'activity', which would suggest that we were excluding non-productive engagements. By 'engagement' we mean of course any duration of time for which an entity is in a constant state — irrespective of whether this corresponds to an active state or one of idleness.

Working generally from the simplest forms of event description, we can examine levels of increasing complexity:

## I   Temporal complexity

The first type of complexity is the simplest of all, where the extent of the engagement is of a purely temporal form, and makes no further demands on other entities nor reference to how they may be arranged. Most commonly this situation applies because all the conditions for progress to the next stage of the entity's process, expressed for instance in terms of resource requirement, have been satisfied except for the mere passage of time. Usually such a situation corresponds to waiting for the occurrence of a prescheduled end-of-activity event.

Sometimes an engagement is restrained until a certain time has occurred. For instance, business may be pursued only during office hours. The restraint is lifted by the occurrence of an event, at a predetermined point in time, corresponding to the opening of the office. In other words, perhaps for conventional reasons, some activities are allowed only during certain periods of time, within the day, week or year. Similarly, any engagement which has already started, and has therefore already been motivated, and whose time-duration is known, is simply awaiting its end-event.

Another common use of purely temporal constraints in a simulation is where we choose to ignore the motivational aspect of an occurrence, such as when dealing with the arrival of entities into a system. In the supermarket model we might, and often do, decide that a sufficient description of customer arrivals would be the assumption of an average inter-arrival rate, randomised according to the exponential distribution. The social and psychological factors which impel individual customers to enter the supermarket are thus regarded as outside the model boundary.

Since the Petri net is primarily concerned with the state dependencies which impede token flow, and has no means of expressing time-elapse, a special means has to be invented to denote temporal dependency, or *delay* (Törn, 1981). We choose the star symbol which permits activation of a transition by a future-event occurrence (Figs 6.4, 6.5). The time advance for discrete-time systems can also be thought of as a series of temporal events, since each time step is determined arbitrarily with respect to happenings in the model.

## II   Configurational complexity

Some events are dependent not upon the passage of time, but upon the occurrence of a certain model-state. The event may be conditioned upon the assumption of a state by a single entity, or, more generally, by a *configuration*, a conjunction of many entities-in-state. Although such state-dependent events refer to entities, the entities comprising the configuration are not necessarily affected by the occurrence of the event. The temporal and configurational forms of complexity, or more simply time- and state-dependent events, thus form the core of simulation representation (Nance, 1980, 1981).

A simple example of *configurational* complexity is the case where a large ship cannot enter harbour until high-tide. When high-tide is the only restraint, its occurrence enables the ship to proceed into the harbour, but it does not mean that the ship 'takes over' the high-tide as if it were a resource. In no sense is there an exclusive collaboration or mutual interaction between the ship and the tide. In fact, high-tide can be shared by any other ships in the locality. The types of configuration upon which events may be predicated are, of course, quite general, and may involve several entities. We need therefore a means of expressing dependency on the occupancy of places, and the transitions they enable (Fig. 7.1). To do this we can make use of the *fact*

Affected
transitions                                                         Places

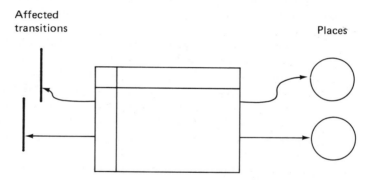

Fig. 7.1 — A fact connects places with the transitions they affect. No tokens flow through facts.

(Reisig, 1985) to connect a dependent transition to places which may affect it (Fig. 7.2).

### III   Resource complexity

The commonest form of entity interaction is one where entities of one type (flowing entities, e.g. customers) wait for service from entities of another type (resources, e.g. servers). The customers are therefore dependent on the servers beings in an 'available' state. But *resource* complexity goes further than configurational dependency in that besides there being a state-dependency, there is an implied entity *interaction* when the configuration occurs.

In many cases, it is sufficient simply to model the resource-limitation by

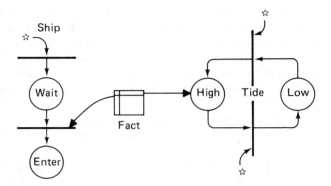

Fig. 7.2 — A ship enters harbour on high-tide. The fact arcs connect the attainment
of high-tide with the ship's entering the harbour.

keeping a count of the customers queueing, and the number of servers idle. Each new entrant into the queue increments the queue length by one, and each removal from the queue for service decrements the same counter by one. We can make the count symmetric between servers and customers by allowing a negative value to represent the number of idle servers. The queue counter is then a measure of the build-up of the traffic at the queueing bottleneck.

The implementation of such a count is obvious; the customer cannot receive immediate service unless the count is at most zero, and, from the server's point of view, customers may only be removed when the count is positive. Since nothing is known of the individuals partaking in the interaction, there can be no detailed representation of either members of the queue or servers. In visual simulation, the count is a convenient way of representing the current state of a bottleneck on a screen. Queue-counters, similar to integer semaphores (Andrews and Schneider, 1983), represent the occupancy of the engagement, in terms of entities awaiting the state upon which passage to the next engagement (e.g. service) can be predicated.

Although a count may adequately represent the queueing system for some situations, the modeller is deprived of the ability to model any idiosyncrasies of individual customers or servers. Expressing entity and resource properties as values assumed by attributes requires a data structure to be declared for each type. It is with the possession of attributes that discrete-event simulation comes into its own and complexity really starts. No longer can a single integer value adequately represent the state of the system, but a full list-processing system with dynamic storage allocation and de-allocation becomes necessary.

Not only is the resulting representation more descriptive, allowing more details to be included, but a wider range of phenomena is possible. For instance, the effect of queueing disciplines, determining the order in which customers receive service, can be easily implemented by defining operations on the pointers which connect the groups of queueing entities.

Normally, we think of queueing discipline as being fixed by the time of entry of the entity into the queue. Thus selecting the customer at the head of the queue, who will have had the *earliest* entry time, will represent a 'first-in, first-out' discipline. Similarly, choosing the customer at the tail, i.e. the most recent entrant, gives 'last-in, first-out' discipline. These mechanisms are *implicit* in the sense that they do not rely on predefined attributes, but only on the order of entry into the queue. Other preference schemes may be based on a *priority* function, defined in terms of attribute values. Instead of joining the queue at the tail, an entity may be assigned its position on the basis of its priority. Where such a scheme might lead to tied priority values, an implicit discipline may be invoked to break the tie.

Simulating each individual queueing entity as a separate process instance enables a qualitative study of queueing models beyond the reach of mathematical modelling. The modelling of queues is widespread in simulation, and it is often regarded as a fundamental form of entity interaction, causing langauge constructs to be devoted to queue representation.

### IV   Select complexity

The model of queueing which we have adopted up to now is based upon that adopted in queueing theory, where the roles of customer and server are sharply demarcated, the customer having a fleeting existence, and the server being stuck on the treadmill of a busy–idle cycle. But this does not exhaust all the possibilities of the ways in which entities can come together.

For modelling general interactions, there is no reason why the entities should be of two different types, or that their number available for interaction be restricted. Nor should it really matter whether their lifetimes are long or short. Conventional queueing normally assumes one entity is served by the other, and therefore has to queue for the privilege, but there is no reason why we should not have general unordered sets for both (or all) types of entity which are candidates for the ensuing activity.

The queueing model can be further generalised by considering the customer and server entities more as equal partners in a mutual *selection*. The 'becoming available' of entities causes a scan or search to be carried out throughout the set of unoccupied entities, to find and possibly initiate a new engagement.

Rather than relying on a queue discipline to determine the interaction, any search criterion may be used, based either on explicit attribute values of the entities, or on implicit duration-of-waiting information, or on a mixture of both. A search criterion may even be one of the attributes of a type of entity; for example, first-fit or best-fit criteria may be employed. In addition, where a set of entities is to act in collaboration, the attributes of already selected entities may also contribute to the selection criterion.

For such a general selection scheme, the waiting entities cannot be ordered for service in advance of the selection taking place. The general program of a *select* transition being in terms of a search means that the control structures for representing this type of complexity resemble those of *query-languages*. All forms of resource–entity relationship, in which entities are passively awaiting the availability of active resources, can be seen as special cases of *select* complexity.

This level of complexity can also incorporate those forms of decision-making common in human and artificial-intelligence studies (AI) in planning, which take into consideration a short presimulation of what will happen in the immediate future, before committing any entity. However, in discrete-event simulation, as opposed to the situation in AI planning, we maintain a strict distinction between instantaneous events and activities with duration, so any decision which takes a significant amount of time, or which cannot be assumed to take place alongside other activities, must be counted as an activity.

In summary, our survey of linear complexities, where we assume only that entities do not stray from a single thread of development, takes us from the temporal event occurrence, which underlies the whole of discrete-event simulation, up to the most general form of mutual selection between many entities. We can observe an approximate correspondence between our linear complexities and the succession of strategies: ES, AS, PI.

As one might expect, the ES strategy, built around the concept of temporal event-driven computation, corresponds to *temporal* complexity where every module is invoked by being scheduled by another module. Any state-dependency must be programmed in by the simulationist. Once we explicitly introduce the concept of 'state', in no matter how simple a fashion, or more generally the model *configuration,* we can consider state-dependent events. The activity-scan of the AS strategy allows any dependency on the *configuration* to be expressed, in the Boolean expression of the C-activity test-heads.

By considering the implications of an entity's individual sequence of activities and its own interaction with other entities — in other words adopting the approach of the PI strategy — enables the representation of *resource* complexity, through the ACQ–REL pair.

Extrapolating this correspondence further, what sort of strategy might there be to supersede the PI? Starting from an analogy with *select* complexity, it would have no prior bias towards the entities presenting themselves for advancement and the requirements of each potential entity would be taken equally into account, especially with regard to contention between rival candidates, etc. This, in fact, forms the basis of the engagement (ENG) strategy, to be introduced in the next chapter.

Essentially, the complexities we have been concerned with so far have been more or less complicated ways of bringing entities together; but the entities have been assumed to be running on tramlines, in that their futures are ultimately fixed and invariable. In other words, we have assumed that the path through the interactions has been *linear*; we now consider the consequences of the possiblity of deviations from linearity.

## 7.3  NON-LINEAR COMPLEXITY

Having dealt with the linear complexities, which as we have seen involve mainly complications concerning the way in which entities are prepared or selected for collaboration, we turn now to the non-linear complexities in which more than one thread of development is required to properly capture the full story of the entity's behaviour. This second source of complexity is a departure from strict linearity in the causation sequence of an entity. Once again, the avoidance of subterfuges is our paramount concern, especially when we would have to invent fictitious entities for the true representation of the system. Unless the simulation language can support such fictions, the simulationist can easily mishandle them, resulting in a false model.

### V   Route complexity

We have already mentioned the central difficulty of simulation, that of representing the parallel development of a set of process instances (entities). Actually, it is a considerable over-simplification to talk of the 'parallel' activation of entities, as if they all proceeded in straight lines. The actual route taken by an entity through a system may involve choosing from a set of possible futures, or maybe there are several lines of development occurring concurrently, splitting and rejoining with themselves, or temporarily submitting to the process of another entity, and subsequently breaking away from it. Many topologies are possible.

The simplest model with route complexity is an example of OR-parallelism, where a future activity is assigned to an entity on the basis of a simple decision, as depicted in Fig. 7.3a. A choice between alternatives may

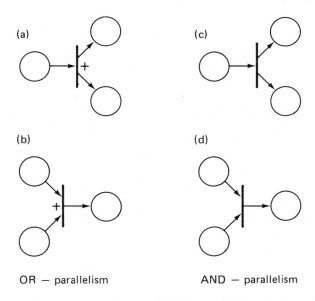

Fig. 7.3 — Simple OR and AND transitions for *route* complexity: (a) at the transition, an output place is selected; (b) a token on either input place will cause the transition to fire; (c) the enabling token is duplicated to occupy both output places; (d) tokens on the input places are matched before firing.

be made on the basis of attribute values, or by a simple probability sampling scheme. Notice that we are concerned here with the choice of possible *futures* for the entity, rather than the entity selection which precdes an engagement. Just as branches can be selected with OR, there are also situations in which *all* of a set of branches must be taken, corresponding to AND-parallelism, Fig. 7.3.

Frequently such situations are met where an entity must acquire a number of different resources before proceeding. Many PI-strategy languages assume, by default, that a manifold acquisition is carried out in an incremental manner, i.e. as soon as one resource of the collection becomes available it is reserved for the collaboration, then the next one is demanded, and so on, until the complete set is obtained. The problem with this assumption is that in reality it is less likely that entities will be reserved when there is nothing for them immediately to do. Moreover incremental acquisition will tend to introduce an artefact into the simulation by requiring a sequential relationship between the entities being selected, where no such relationship actually exists.

Languages based on the process concept tend to encourage sequential acquisition of resources, even when what is meant is a representation of an all-or-nothing type of acquisition. Incremental acquisition can often lead the modeller unwittingly to describe a *deadlock* situation, in which two entities are unable to progress as each has got, but cannot yet release, what the other wants. An AND-parallel means of specification could ameliorate this problem, by enabling individual acquisitions to proceed simultaneously. The two means of specification for multi-aquisition are compared in Fig. 7.4.

In the case where one entity *submits* to another, the appropriate type of interaction is that of a temporary suspension of the submitting process, which may be resumed subsequently by the original entity. The familiar customer–server interaction is of this type. Apart from the 'idle' state, the states of the server are determined by the customer's process.

For another example, consider a lift-passenger waiting for a lift to arrive which can take him to his destination floor. On the arrival of the lift, the passenger gets in and is, from then on, subject to the stopping and starting of the lift as it deals with other passengers getting in and out. During the interaction with the lift, the passenger's states are determined by the lift's activities. On arrival at the destination, the passenger gets out and may resume his own process instance from where it was left off.

Whenever the object system requires the modeller to override the resource–entity distinction, which is strictly enforced by most PI languages, we are forced to enter into a subterfuge. Conventionally we would see the passenger (a 'temporary entity') as the customer and the lift (a 'permanent entity') as the server, but this would require some subterfuge in order to model the sharing of the lift between different passengers, as servers are usually regarded as being exclusively bound to their customers during the 'service' engagement. As we have already emphasised, the danger with such departures from reality is that an unforeseen interrupt during this period would receive a false impression of the model state.

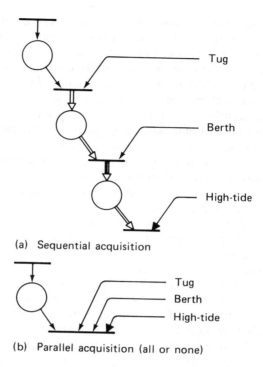

(a)  Sequential acquisition

(b)  Parallel acquisition (all or none)

Fig. 7.4 — Multi-acquisition specifications to represent a ship entering harbour.

## VI   Virtual complexity

In simulation it is normally taken as axiomatic that upon every engagement creation we may associate an event corresponding to the finish of the engagement, even if we are unsure of its occurrence time. Normally, apart from durations of waiting idle, each engagement may be wound up by a specific event occurrence, called the *outcome* of the engagement. One question to consider is, can we give any meaning to situations where we may have more than one outcome?

It would be unwise to allow a multi-outcome to represent a sequence of future happenings for an engagement, because allowing assumptions to be made about events happening too far ahead in the future is unwise, as locally unforeseen events may occur to forestall them. Were we to allow an engagement several outcomes, then it seems natural that whichever is the *earliest to occur* should be the outcome to dissolve the interaction, the other outcomes being cancelled on its occurrence. Thus a *multi-outcome* engagement is one in which several outcomes are set up simultaneously, each one entered *virtually,* indicating that the engagement may be terminated by any one of the set of events occurring.

A simple example of a multi-outcome is familiar to users of time-sharing computer systems: if a certain time elapses between successive terminal accesses then the user of the terminal is 'timed-out'. The computer session is thus terminated by one of two virtual outcomes: normal finish, or exceeding the allowed period without touching the keyboard.

For a more substantial example of virtual outcomes, we may turn again to lift-usage. Consider how to model the impatient lift-user: someone in a hurry is trying to get to another floor in a multi-storey building. He decides to wait for the lift, with the proviso that, if it has not arrived within a certain period he will give up his wait and walk to his destination. So, at the start of the wait, the person's future is decided by the earlier of two outcomes: either the lift arrives and he gets in, or his patience is exhausted and he walks to his destination.

The impatient lift-user can be represented (see Fig. 7.5) as two aspects of the passenger's behaviour occupying a linked place. Each aspect awaits a separate outcome: patience exhausted or lift arrival. On the occurrence of

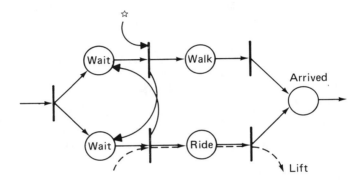

Fig. 7.5 — Impatient lift-user. On entry the passenger is represented by two aspects: one awaiting the end of the passenger's patience, the other the arrival of the lift. Since it is not known which will occur first, the two aspect-tokens reside in a linked place called 'wait'. When one of the outcomes occurs, the other aspect-token must be destroyed. This connection is shown by ......▶ denoting a flow of information.

either outcome, the potential occurrence of the other must be disabled, which involves either the deletion of an FE notice from the FES or the deletion of an element from a waiting set. This is denoted in the diagram by information arcs connecting the critical transitions with the appropriate place of the *opposite* aspect. If the language recognises the possibility of virtual outcomes, then this cancellation can be carried out automatically. Frequently the parallelism may extend further than a single engagement, and the virtual parallel futures may comprise separate processes. In this case there must be a means of specifying the point in each process at which the other outcomes become invalid.

Another example of a pair of virtual outcomes occurs when simulating with continuous flows, which remain at constant rates between discrete events. Consider a vessel being filled, with a certain level of input and output flow. If the net input rate exceeds net output then the state 'full' will eventually be reached, at a time which may be predicted. However, such a prediction may be invalidated if there is the slightest change in flow-rates, even to the extent of possibly changing the outcome from 'full' to 'empty'. During the activity, the engagement is associated with a multi-outcome, the full and empty outcomes remaining virtual right up until the occurrence of one of them.

To implement such effects as submission, multi-entity collaboration and virtual outcomes, the *engagement* data structure (Fig. 7.6) has been put

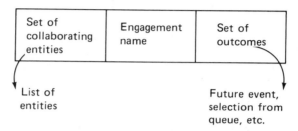

Fig. 7.6 — The engagement data structure. From Evans (1981).

forward (Evans, 1981) to link together all relevant pieces of information that might be needed to fully represent the potentialities of the current activity. An engagement persists for the duration of the activity, and serves as a complete descriptor of the activity during its simulation, by linking together three pieces of information:

- name of current state;
- set of entities currently collaborating in this state;
- set of outcomes which could bring this state to an end.

### VII   Interrupt complexity

In the case of virtual complexity we have seen the extra modelling capability afforded by slackening the inevitability of the process sequence and allowing an engagement more than one outcome. The virtual outcomes are all predictable to the extent that their possible occurrences will have a specific expected effect on the current state, but there is uncertainty as to which will happen first.

We can extend this idea by allowing a physical entity's behaviour to be described as a multiple of *quasi-independent* processes, representing different *aspects* of the entity's behaviour, each capable of frustrating the sequential implications of another. The separate aspects evolve concurrently, and the interactions between them occur as unexpected *interrupts* upon the subject entity.

Perhaps the most commonly discussed situation where interrupts arise is when a discrete model and a continuous model are combined. Because each type of simulation uses a different approach to time advance, we need a means of communication of events between the separate parts of the model. In this way the effect of threshold levels being attained in the continuous part may be relayed to the discrete part, and events happening in the discrete part may exert an influence on the continuous part.

The assumption of relative independence between aspects communicating through interrupts considerably facilitates the modelling of a complex system. Interaction between aspects through interrupts enables one aspect to commandeer a subject aspect and change its course of action. Moreover, the precise nature of the effects of the interrupt may be made dependent upon the state of the subject aspect on interrupt occurrence.

One of the commonest applications of interrupts is the simulation of *breakdown*, i.e. a sudden, drastic failure in a component of the system. Such events are unexpected and often catastrophic in reality, demanding emergency procedures to be put into practice to alleviate the consequences. As such they are excellent candidates for computer simulation, to enable the effectiveness of emergency procedures to be safely gauged. Their modelling difficulty lies in the reneging of some future events, and the temporary following of an alternative process.

Interrupts can often be caused by an accumulation of time in a particular state, or set of states, or of the accumulation or depletion of a flowing substance. As examples we may cite the ending of a period of a machine's working-life, or storages reaching the limit of their capacity. Normally, with a discrete-event system, one assumes that between reallocations at model state-changes everything remains unchanging. However, such an assumption can be mediated in the case where a flow is a linear function of elapsed time; for although flow-rate may be constant, accumulation may continue and critical levels may be attained as a result. In a discrete model we may allow linear flows whose rates change only at state-changes. Allowing continuously changing rates would of course amount to a continuous model.

Let us consider the occurrence of a mechanical breakdown determined by an accumulation of working hours. Breakdowns often occur owing to the wearing out of critical machine components which in turn happens at a rate proportional to the exposure of the machine to work. Thus the machine is modelled as a pair of aspects, one representing normal working, the other the wearing out of an important component. In the model, exposure to work is represented by the adoption by an entity of specific states, so to gauge the next breakdown event implies the accumulation of time-in-state, for the one (or more) working state(s).

Breakdowns may have many causes, and many effects. The effects are often dependent on the state of the subject entity at the time of the breakdown, i.e. at interrupt occurrence. If the entity is in a state of idleness, then the effect of the interrupt may be slight, or even ignorable. Alternatively, its occurrence may not be immediately detected, but the entity may behave faultily when next pressed into action. Most probably more extreme consequences will occur when the entity is in an intensive activity, working in collaboration with other entities. Thus the effect of a breakdown in one component may propagate to other entities. Simulation would be an incomplete technique without some facility to model an entity as a multiple of quasi-independent aspects.

The overall picture of a breakdown simulation, or any similarly cata-strophic occurrence, can be resolved into three sub-simulations:

- the normally functioning system;
- the cause of the failure;
- the emergency procedures.

With Tocher (1969), we take the optimistic view that catastrophes can be ignored until they happen. It is more convenient, to say the least, to write the main simulation as if it were always functioning normally. Otherwise including preparation for the possibility of breakdown could well cause each process to be hideously complicated, bristling with conditional statements, since almost any happening in the simulation could have an effect. In doing so we separate the breakdown and normally functioning aspects of machine behaviour.

For the total lifetime of an entity-aspect, the state and all its associated details are collected in the engagement. Every state-change undergone by an entity-aspect is modelled as a new engagement instance, and progress through a process can be seen as a succession of engagements, alternately coming into and going out of existence. The facilitation of interrupt modelling is the central reason why we introduce the idea of naming each engagement within a process.

On the occurrence of an interrupt, the engagement data structure provides the source of information for the actions to be effected: the current activity field enables state-dependent effects, defined in the interrupt module, to be selected for activation. Propagation of the effects may be channelled through the field linking the collaborating entities, and the set of outcomes, to cover the case of a multi-outcome engagement, enables previously scheduled events to be cancelled, and perhaps rescheduled.

Besides breakdowns, interrupts may arise from other interactions between models which are quasi-independent, and may be regarded as a communication through message-passing. For instance, it may be beneficial to regard the relationship between a warehouse and its supplying manufac-turing plant as two separate simulations interacting by interrupt. Moreover, systems can often be defined at different levels of interest. Usually, traffic flow along a highway may be regarded as some kind of continuously flowing entity, until it becomes necessary to simulate the behaviours of individual drivers, which may involve a more discrete model, perhaps involving some presimulation to account for drivers' interpretations of the intentions of their fellow road-users.

Although the simulation of interrupts is not at all common, due probably to the poor support offered by simulation languages, the methodology of integrating partial models to simulate a wider complex whole, as suggested by Zeigler (1984), may lead to a wider application of interrupts, mediated through engagements.

## 7.4 COMPLEXITY IN SIMULATION LANGUAGES

The aim of the simulationist is primarily to accurately represent a dynamic reality by the flow of control in a computer program, upon which he can reliably experiment without interfering with the real system. It follows that if the language he uses is to be at all comprehensive, then it must be able to support many patterns of control flow, enough to match a wide range of dynamic behaviours with which the simulationist is likely to come into contact.

We are familiar with the control structures offered by conventional languages, such as Fortran and Pascal. They may be grouped into three broad categories:

(i) conditional branching;
(ii) iteration;
(iii) subprogramming.

All categories involve a sectioning of the program, so that sequences of commands may be treated as wholes, to be jumped around or to be executed *en bloc,* as determined by local factors. In addition the sectioning will impose a *data control logic* (Marlin, 1980) over the program governing the visibility and lifetime of the variables which are declared within each section. Normally we find that there is a hierarchy of variables, corresponding to the hierarchy of sections, giving rise to the familiar pattern of a block-structured program. In the case of the interrupt, for which the relationship between the behaviour of one aspect is completely arbitrary with respect to another, a hierarchical relationship is obviously insufficient. The question arises as to what extent these complexities can be interpreted by the control structures of simulation languages. We shall now survey three common simulation languages, GPSS, SIMSCRIPT and DEMOS, with regard to this question.

### *I   Temporal complexity*

GPSS:         We have the GENERATE statement (Gordon, 1975) to schedule the birth or arrival of new transactions into the model. Their attributes are also defined in the same statement. ADVANCE is used to denote the passage of a known duration of time, which may be specified as either a constant or a distribution from which a sample is to be taken.

SIMSCRIPT: The passage of time is indicated by either of the synonyms WORK or WAIT. Generation of new entities requires the modeller to define a generator process, usually in the form of an ACTIVATE statement executed from within a loop defining the inter-arrival properties of the entities. (Russell, 1983, pp. 2–12).

DEMOS:     Following its parent Simula, to indicate the passage of time, DEMOS (Birtwistle, 1979) uses the *hold* statement which applies to the entity at the head of the event list, and *schedule* can be used to generate new entities at a particular inter-arrival rate. Like SIMSCRIPT, there is no special statement to generate arrivals.

### II Configurational complexity

GPSS:　　　No direct support for configuration detection is offered, but each block has associated with it a list of so-called standard numerical attributes (SNAs) which are accessible from any-where else in the program. The information these attributes contain refers to current status (e.g. in use, or average length of queue) and past history (e.g. total number of transactions entered since simulation start). Thus a wide variety of configurations can be identified, and their occurrence can be detected by the conditional statement TEST. The responsibility for detection lies entirely with the simulationist, of course.

SIMSCRIPT:　Analogous facilities to those of GPSS are offered.

DEMOS:　　The synchronisation pair *waituntil* and *signal* can detect the occurrence of a configuration, as long as the simulationist includes a 'wake-up' *signal* call at each location where the configuration might turn out to be **true.** The argument whether or not to automatically include a search for awaited configurations is fully discussed by Birtwistle (1979), who decides it to be the user's responsibility to reawaken passive entities. These points will be discussed further in Chapter 8.

### III Resource complexity

GPSS:　　　Acquisition of resources in GPSS is divided into two types, depending on whether the resource is a *facility* (i.e. there can only be one instance of this type) or a *storage*. The SEIZE block is used to acquire facilities and when none is available, transactions are queued FIFO to await satisfaction of the acquisition. For storages, any amount of resource can be requested at an ENTER block, but the default queueing arrangement favours transations with a lesser requirement. More exotic disciplines using LIFO or based on priority schemes may be set up with *user chains* accessed through the LINK/UNLINK pair.

SIMSCRIPT:　REQUEST/RELINQUISH is the ACQ/REL pair for conventional entity–resource interaction. SETs, or an ordered collection of entities, are used to model queues. Queue disciplines are FIFO by default, or may be LIFO or RANKED according to an attribute. A priority queue may be modelled either by a ranked SET or by using a REQUEST WITH PRIORITY statement.

DEMOS:　　For queueing for a standard resource in a busy-idle cycle, we have the RES entity, with the pair of synchronisations

ACQUIRE/RELEASE. ACQUIRE may be parameterised by the number of resources required. By default, if there are not enough resources to meet the demand, the entity enters a FIFO queue.

Alternatively, the resource may be modelled as a BIN, with synchronisation pair TAKE/GIVE, the difference being that an entity of type BIN has no upper limit on the number of items, whereas RES has a fixed number of instances, as set out in its initialisation statement. Basically both forms of resource are modelled simply as a count of entities. Support for LIFO or priority-based disciplines appears to have been omitted from the language.

## IV   Select complexity

GPSS:   In order to carry out a selection from a set of transactions, the set must be LINKed into a user chain, and every time a selection is required a fictitious iterating transaction is brought into play to execute the search loop. Transactions in the chain can be selected and removed with UNLINK, on the basis of their attribute values, but each attribute-condition requires a separate user chain. The resulting program can become hideously complicated, and it is difficult to get right. This aspect of GPSS is probably its weakest.

SIMSCRIPT:   The 'FOR EACH *resource*' statement causes a scan over the named resource set in indexed order. Further selectivity can be added with a 'WITH X .R. Y' condition, where X identifies a parameter and Y value, with .R. a relation. Such conditions may be further ANDed together. A FIND statement can be used to terminate the loop after the first resource satisfying the condition is found. The result of the search can be ascertained by the inclusion of an IF FOUND or IF NONE clause after the FIND statement. A full discussion, covering looping, selection and searching, is given in Maryanski (1980). This is a comprehensive, if somewhat muddled, facility, and applies to resources only.

DEMOS:   FIND locates a suitable entity for collaboration from a particular queue and blocks the caller if none is available. The parameters of FIND identify the queue (selection set) from which the selection is made and a **Bool** expression, which is passed by name, is the search criterion.

## V   Route complexity

GPSS:   There is no way in which transactions can explicitly cooperate with one another, except in the case of a resource–entity interaction as discussed above. One obstacle is that references to transaction SNAs are self references — there is no mechanism to refer to the SNAs of another transaction —

thus it is difficult for transactions to process jointly. Some versions of GPSS (e.g. Univac's) do have a means of doing this. However, apart from these shortcomings, we may spawn transactions from a SPLIT block and re-integrate them with ASSEMBLE or GATHER, which provide for matching offspring with the originator, an important system task. Thus we have a capable means of modelling multi-outcome and multi-aspect entities, if handled carefully. Of the widely available simulation languages, only GPSS pays much attention to this.

For branching based on a decision, there is a TRANSFER command which allows a label to be specified which is jumped to on the basis of an evaluated expression.

SIMSCRIPT: Although Russell (1983) claims to support complexity beyond entity–resource, there are few examples of it. There is a description of a simple conditional IF statement to achieve OR-parallelism. CASE statements are also available, but there seems to be no representation of multi-aspects, nor of one entity submitting to another.

DEMOS: Entity–entity interactions through COOPT/SCHEDULE (0.0) are specifically allowed (Birtwistle, 1979, p. 76). An entity entered into a *waitq* can be associated with another through a call of COOPT. At the end of the cooperation the entities may separate from each other with a call of SCHE-DULE (0.0). No splitting of a single entity is supported.

## VI   *Virtual complexity*

GPSS: Virtual aspects arising from multi-outcome engagements can be modelled using the subterfuge of a SPLIT transaction. The modeller must remember to include the termination of the transaction whose outcome did not occur. An example of time-limited waiting (the impatient lift-user) is in Gordon (1975, p. 293).

SIMSCRIPT: Apart from being able to CANCEL previously scheduled events, there is little provision for virtual outcomes.

DEMOS: Similar to SIMSCRIPT, there is a cancellation routine, which is an attribute of the CLASS ENTITY, but little else besides. It may be used to remove future-event notices from the FES, or to terminate a simulation run prematurely.

## VII   *Interrupt complexity*

GPSS: The PREEMPT/RETURN synchronisations allow a simple kind of interruption of a facility, whereby a PREEMPT will override a previously issued SEIZE command, and the pre-empted transactions will be put into interrupt status. On RETURN the original state of affairs can be regained, if desired. Access is also offered to the remaining time that the transaction would have spent in the block if it had not been interrupted, enabling the original activity to be resumed for the period outstanding when the interrupt occurred.

Sometimes it is necessary to further interrupt a transaction which itself has already interrupted another transaction. On RETURN either a LIFO resumption of control or a resumption with respect to a priority scheme may take place. Other operand settings allow simulation of queue-jumping by replacing queued transactions of lower priority.

GPSS V enables resources to change their availability by arranging control transactions to flow through AVAIL blocks at appropriate times. The AVAIL blocks are FAVAIL and FUNAVAIL for facilities and SAVAIL and SUNAVAIL for storages. The primary purpose is for modelling breakdowns (Gordon, 1975, p. 98). Their functioning is quite complicated. For instance, FUNAVAIL has up to eight parameters because it has to deal with three classes of transaction: the one using the facility, the one that has been pre-empted, and the ones waiting their turn to use the facility.

The lack of any means to cancel previously scheduled future events is a serious handicap in more general interrupt modelling. Another drawback from using GPSS at this level of complexity is its lack of clarity, with important details getting lost in a maze of block attributes. The very conciseness of expression which is attractive in simple models becomes obscuring and awkward with increasing complexity.

SIMSCRIPT: To obtain the effect of PREEMPT of GPSS, SIMSCRIPT has an INTERRUPT (Russell, 1983, pp. 3–31) command to displace a process on the pending list (i.e. the process must be known to be currently involved in a WORK or WAIT statement). Elapsed time remaining at the occurrence of the interrupt is obtainable as a process attribute (TIME.A). On subsequent resumption (by a RESUME statement) this value is used to determine the new duration of work.

Processes may be prematurely terminated by being subject to a DESTROY statement, normal termination occurring through either RETURN or END. The interrupt capability of SIMSCRIPT seems to be about as comprehensive as that of GPSS. However, the simulationist must take care that the interrupted process is in the pending list, which might be awkward to arrange, since processes awaiting resource availability cannot be interrupted. However, the cause of an interrupt can be effectively modelled by the occurrence of an exogenous event, or the development of an exogenous process.

A combined discrete–continuous version of SIMSCRIPT

DEMOS:

called C-SIMSCRIPT has been developed (Delfosse, 1976). Some simple interrupt features are included, without attempting to be comprehensive (Birtwistle, 1979, p. 122). A kind of GPSS-style PREEMPTion example of lathe break-down is given, with interruption of work to be resumed for the remainder of the work period at a later stage. This can be carried out using CANCEL/SCHEDULE. When interruption implies loss of collaborating entities, to be recovered in competition with other entities' demands on resumption, we use the procedure INTERRUPT which is another attribute of CLASS ENTITY. Different interrupt actions may be specified by a distinguishing integer parameter. But there is no way of having the interrupting actions dependent on the current state of the interrupted entity.

Taking advantage of the package-writing capability of Simula, several combined languages have emerged: DISCO (Helsgaun, 1980 and CADSIM (Sim, 1975).

### Other languages

The first paradigm implies that all simulation languages must be able to handle temporal events. But the concentration of concern with state-changes occurring at predicted times has led to a strange variety of synonyms to describe the same function: hold, cause, wait, advance, time, etc.

The effect of strategy soon shows itself when we consider configurational events, for which AS languages are especially adept at describing. The PI strategy offers the simplicity of the resource assumption. As for selection, the languages which initially saw the simulation problem as one of handling sets, namely SIMSCRIPT and CSL and its derivatives, fare better than the others.

Outside GPSS, little attention has been paid to simulating with multiple aspects. The language GERTS (Elmaghraby, 1977), which is designed for modelling *activity nets*, and therefore concentrates on the representation of probabilistic branching and various kinds of *route* complexity, includes a virtual branch feature which goes against the general assumption in network processing that one always has to wait for the most delayed predecessor before proceeding with an activity. With this feature we await the earliest of a set of predecessors.

Considering languages' approach to modelling interrupts, we should mention the AS language HOCUS (Poole and Szymankiewicz, 1977), in which 'split' entities (aspects) can be involved in several continuous activities simultaneously. A discrete activity can be used to interrupt a continuous activity, to model the operations of, say, chemical plant, with tanks filling and emptying, feeding into one another, with constant flow-rates. Further-more, such entities may break down and their end-of-activity events may be re-scheduled.

One of the first papers to present an example of a system involving continuous activities and discrete events capable of mutually interrupting one another is that of Hurst and Pritsker (1973). The language GASP IV is used. Hooper and Reilly (1983) have proposed a combination of GPSS and GASP to provide more support for combined continuous and discrete-event simulation. A combined simulation AS executive has been proposed (Ellison, 1979), where the continuous thresholds are incorporated into extended C-activities. A survey of combined simulation software is given by Ören (1977).

One rarely considered, but quite important, stage in a simulation run is *initialisation*. The operations involved are essentially mapping entities into activities, or tokens into places, which is just the kind of operation we require to model the effect of catastrophic changes. We will consider this further in Chapter 8. Many of the concerns of simulating complex systems are similar to features of other kinds of programming, especially when programs are to run concurrently. Some of these analogies are considered in the next section.

## 7.5 SIMULATION AND CONCURRENT LANGUAGES

### 7.5.1 Interrupt as a control structure

We have seen that for non-linear forms of complexity we must be concerned with multiple sequences of activities executing, at least notionally, in parallel. A sequence corresponds to the progression of a single aspect of an entity's behaviour, whether the aspect is real, potential or virtual. The types of interaction between aspects may range from being largely independent, through being dependent on an expected occurrence in a brother aspect, to being prone to interruption at a critical occurrence. The more the variety of interactions, the more complexity (Riddle and Sayler, 1979) can be modelled. From the language-design point of view, the interrupt, being a signal from the environment external to the part of the program currently being obeyed (Barron, 1977), is the most difficult situation to support.

The types of complexity discussed above may be compared with the control structures supported by concurrent programming languages, which also have a need for separate processes to synchronise, communicate and interact with one another. In concurrent programming the term 'interrupt' is used more generally than in simulation. For instance, exception handling, which is usually regarded as a form of interruption in concurrent languages, would often correspond to a configurational or select complexity in simulation.

The idea of an interrupt as a break in the normal control sequence of a computer's running program is as old as computing itself. It appealed to Babbage for his Analytic Engine and has continued to find a prominent place in the control of operations at the hardware and operating system levels. Indeed the notion of interrupts is so engrained that many operating systems are interrupt-driven (Kuck, 1978). Thus whenever a significant change in the control state of the computer arises, an interrupt is generated and the operating system takes over control, at least momentarily. In this way a great diversity of conditions which may arise can be handled in a uniform way.

Yet despite its importance at low levels of computer operation, it is rare to find any provision for interrupts as a high-level programming tool (Barron, 1977). PL/I is the only commonly used language to recognise its existence. The reason for this is probably the interrupt's maverick nature, which it shares with the **goto** (Dijkstra, 1968b), in its ability to cause drastic changes in the flow of control, and render the resulting program too intricate to be easily understood. Both control structures go against the tenor of structured programming which emphasises a more circumscribed approach to control flow. Dijkstra (1976) reflects that much of the work on cooperating sequential processes has been for the purpose of taming the interrupt, which was regarded as a curse inflicted by the engineers on the software designers.

For the lowest level of language, i.e. machine language, control structure is limited to the **goto** and interrupt. The types of interrupt encountered in the hardware–software interface are manifold. Garside (1980) has identified ten distinct causes of interrupt. Some interrupt classes are more urgent than others and require a priority scheme to be effected. Their unpredictable occurrence means that some strategy has to be adopted when two or more require attention at the same time, or when one happens during the operation of the effects of another which occurred a short time previously. Thus the possibility arises of having to store the part-results of one interrupt operation while a more important one is attended to. The particular policies adopted regarding priority and pre-emption issues give each operating system its own individual flavour.

To some extent, interrupts are antagonistic to a clean program structure. Yet interrupts abound in designed systems. The telephone may have been intended for long-distance audio communication, but in use it is largely effective as an interrupt machine. Any facility to model interrupts must be able to transcend the rigidities of 'good' structure in order to render a precise representation.

### 7.5.2  Operating systems

It is probably true to say that the most widespread use of interrupts is in the field of input–output programming. If we go back to the earliest stored-program computers, they suffered from the major weakness that the arithmetic–logical unit could be seriously impeded by the much slower processes of input–output (Enslow, 1974). To get round this problem, an input–output channel was installed to provide a separate access path to memory and a control path dedicated solely to input–output. While the input–output channel control was limited in its actions, its simultaneous running and parallel relationship with the main processor greatly improved performance.

However, in order to keep the control paths in step with one another, a special program was necessary which has grown into today's operating system. Whereas a uniprocessor may be programmed directly for an application, even the simplest system with parallelism requires system software. Input–output programs are generated and placed in memory for access by the input–output channel which executes them while main processing continues. When the input–output is complete, an interrupt signal is passed to the main processor, indicating that either the task has been completed or an error has been encountered. Operating systems developed out of the requirements to handle interrupts and such issues as the preservation of data integrity under reading and writing operations. The effect of this creeps through into high-level languages. In Fortran, READ commands can use 'END=*label*' to denote a label to jump to when the end-of-file marker is sensed, a simple example of an exception interrupt, i.e. a wait for an expected occurrence.

For operating systems which perform time-sharing, the running programs are are to be run in parallel. The illusion of parallel running is given to each individual user by allowing each program access to the processor for a succession of time-slices. Advantage is made of the fact that most programs spend a large amount of time waiting for an input–output interrupt, so rather than have them wait in main memory, they are swapped out onto disk. The arrival of the interrupt then serves to signal the re-entry of the job back into main memory. Jobs requiring continuous residence in memory are allocated processing time in small units at the end of which the clock-interrupt signals that another job requiring processing should supplant the current one.

The interaction of ideas between designers of simulation languages and designers of operating systems has been persistent. Brinch Hansen (1977) notes that borrowing the *class* concept from Simula in the development of Concurrent Pascal gave a great deal of insight. Simulation of operating systems, as well as other software systems, such as computer networks, is a good way of testing their effectiveness. The language Concurrent Euclid (Holt, 1983) may be used in simulation mode, with a *busy (duration)* statement to represent simulated elapsed time.

Other specialist simulation languages exist, e.g. APL*DS (Giloi *et al.*, 1978), for simulating hardware components at the register transfer (RT) level. As Birtwistle (1979) remarks, operating systems are conveniently modelled using the PI strategy, because the text of the process closely follows the sequence of the programs. Significantly, the authors of a recent text on VLSI simulation (Hill and Coelho, 1987) also adopt the PI strategy.

What we have in an operating-system program is thus a highly sophisticated piece of software which is activated largely by interrupt occurrence (Dijkstra, 1968a,c). However, the operating system can take advantage of one fact which does not apply in simulation: each individual job, being the responsibility of each separate user, is completely independent of any other. It would be an indication of grave malfunction if the happenings in one job were able to influence another job in the system. The kernel program handles all jobs as fairly as possible and the jobs fall completely under its control while they are in the system, all the time each job remaining independent.

Concurrent languages have a similar difficulty to simulation languages in the representation of interrupts. Wirth (1982) identifies an interrupt as 'a transfer statement into a coroutine at a point which is not previously specifiable', thus recognising the arbitrary nature of interrupt occurrence, and gives an interrupt-handler program for a computer keyboard in Modula-2. For a simulation, the whole gamut of interaction (non-linear complexity) between parallel processes is possible, and should be supported in a comprehensive language.

### 7.5.3   Exception handling and multi-tasking

A software treatment of some varieties of interrupt communication is found in the Burroughs extension to Algol (Organick, 1973). For an interrupt to be defined, a recipient will have designated in advance the procedure that should be executed on receipt of that interrupt signal. If the recipient is not executing, then the signal is re-triggered when the task is reactivated. In cases where the recipent is to be prepared for the interrupt occurrence beforehand, the occurrence can be regarded as an *exception,* and the execution of the occurrence-dependent code as *exception handling.*

In Burroughs Algol (Organick, 1973) for the B5700 and B6700 series, which differs from its predecessors mainly in the consideration of the problems encountered in multi-tasking, interrupt effects are defined as parameterless procedures with **'interrupt'** supplanting the word **'procedure'**. They are attached to events and readied for use as follows:

> **event** *ev*;
> **interrupt** *rupt1*; **begin** . . . **end**;
> **attach** *rupt1* **to** *ev*;
> **enable** *rupt1*;

Thereafter whenever the **event** *ev* appears as the parameter to a call of the system procedure *cause*, the interrupting procedure *rupt1* will be activated.

The Burroughs Algol specification of a set of tasks is that they should run in parallel: whether this is physically realised, or the sequences are somehow interleaved along a single sequence, depends on the implementation of the language and the computer configuration. In principle this should be of no concern to the programmer — the implementation should take advantage of whatever configuration of hardware it finds itself running on. Tasks communicate through the type **event.** Each task has four attributes:

- status,
- exception task,
- exception event and
- partner.

The statuses which a task is allowed to adopt are similar to the states of a Simula process: suspended, activated or terminated. Whenever a change in status occurs in task A, the **event**

*A.exception_task.exception_event*

is *caused,* that is to say the status-change is noted and any task which was waiting for just such an occurrence is reactivated. In this way the effects of a status-change in one task may be broadcast to other tasks that may be waiting for this occurrence. Naturally, contention problems between waiting tasks will arise, and the situation is rather similar to contention between C-activities on the occurrence of a particular B-activity, or between different transition firings on the occurrence of a particular configuration. When two tasks are partners, i.e. if

$$A=B.partner \text{ and } B=A.partner$$

then the two tasks operate as a pair of coroutines.

PL/I allows the programmer to monitor the program throughout execution (Nicholls, 1975), allowing transfer of control to take place when a particular set of circumstances occurs, causing an interrupt. Besides input–output interrupts, PL/I allows several other classes of interrupts (PL/I conditions) to be handled, covering program check-out, storage and system action areas. The PL/I programmer thus has the power to recover the program and to take remedial action in such cases as dividing by zero, undefined file, index outside subscript range, insufficient storage allocated or any system-located error.

From the very nature of these conditions, we can see that the primary purpose of PL/I interrupt handling is as a stop-gap, to decide what to do when something goes wrong. The programmer is able to define his own conditions for interrupt, but since the recipient is always required to be prepared for the interrupt, by being within the context of the appropriate

ON (*condition*);

context, the interrupting agent is hampered in its operation. In many cases we could get the same effect by writing code to explicitly uncover a condition beforehand, but when used carefully this facility can be an effective debugging aid.

This attitude of PL/I towards the interrupt has been largely followed in the design of Ada (Ichbiah *et al.,* 1979). But since Ada is designed with real-time applications strongly to the fore, the interrupt, in the guise of the exception, rather than being 'bolted-on' as in PL/I, has been acknowledged in the general format of the basic program block:

```
DECLARE
-- declarations
BEGIN
-- statements, the body of block
EXCEPTION
-- exception handlers
END;
```

Any exception may be raised in the body of a block by a statement of the form

RAISE exception_name;

whereby execution of the block ceases immediately and control is passed to the exception handler in the block, which is of the form

WHEN exception_name=>
-- sequence of statements;

with the reserved word OTHERS to provide a general exception handler for any exceptions not explicitly handled.

Although Ada pays quite a considerable amount of attention to interrupts, in the form of exceptions, they are in fact designed for a specific type of usage which prevents them from being applicable in the case of simulating machine breakdowns, for instance. This is because the design of the language expressly forbids that control should revert to the point where the exception occurred. The reason for this requirement is probably the wish not to attempt to re-enter a processor in the case of its malfunction. Such a restriction is quite understandable in the context of reliable real-time operation covering many concurrent processors. The exception handler is thus seen as performing a kind of tidying-up exercise, should the block halt for any reason which causes doubt about its future operations.

Moreover, in the definition of the exception used by the Ada designers, Goodenough (1975) deliberately excludes a general-purpose interrupt capability. In this definition of the exception, the invoker of an operation is merely provided with information that a certain condition has come about. The invoker is then permitted (or in some cases required) to respond to the condition. The operation itself is not permitted unilaterally to decide what action is to be taken. For simulation application, we require a more general range of responsibility for interrupt handling.

Both PL/I and Ada contain facilities for multi-tasking. Petrone (1968) proposes a Simula-type language using the multi-tasking features of PL/I. The facilities in Ada are more sophisticated. Ada tasks communicate and synchronise through the *rendezvous* where a task may wait on an ACCEPT statement, or by using the SELECT statement await several alternatives, or take up default actions (Unger *et al.*, 1984).

In Ada, the declaration of a task can specify an entity life-cycle in a process style, and process instances can be generated, etc., but unfortunately the entity's attributes remain hidden to other entities, a property which prevents the expression of any *select* complexity.

For the simulation of general interrupts, the type of inter-process interaction would correspond, in a multi-tasking environment, to the wresting of control from one task by another. Any decision whether or not to return control would be at the option of the superior task. No such inter-task communication is allowed in PL/I, however, and Ada, once again conforming to its remit, allows only the built-in exception FAILURE to be raised in another task:

RAISE *another_task* FAILURE:

which causes the interrupting of the execution of the named task. The occurrence of the exception is noted and execution may continue in the task where the exception was raised. The authors (Ichbiah *et al.*, 1979) place great emphasis on the extremity of the situation in which FAILURE should be used, for instance to protect a task against a possible malfunction in another task, or to terminate an erroneous task. It is a drastic measure, to be used only when normal means of communication have failed. Further problems with Ada simulations are discussed in Chapter 9.

### 7.5.4  Real-time systems

It is in the area of real-time programming that a similar level of complexity arises to that of simulating general interrupts. This is because it is a common requirement, e.g. in a process-control environment, to be able to cause the events occurring in one process to affect others, or to enable events in one process to respond to external stimuli within a finite and specifiable delay (Young, 1982). As in simulation, many special-purpose languages have evolved to assist the programmer in the control and monitoring of processes operating in real time. Both sets of languages have converged somewhat on the common concept of the *event* to express control concepts. In fact Nygaard and Dahl (1978), Gordon (1978) and Tocher (1964) all mention that their languages have been considered for real-time application.

For instance the real-time language RTL/2 (Barnes, 1976) allows the suspension of a task for one of three reasons;

- waiting for a specific delay time;
- waiting for an event to occur;
- waiting for a semaphore to be released.

Each case has an analogue in simulation, corresponding to temporal, configurational and resource complexity, respectively.

But the provision of high-level language concepts in real-time programming does not have so long a history because of the very high premium placed on run-time efficiency, which tends to make programmers write directly in machine code. With many different kinds of processes being brought together involving different manufacturers, the common agreement and acceptance of a single real-time language is more problematic, so much programming remains at the bit-manipulation stage. Ada supplies an adequate set of real-time facilities (Gehani, 1984), including the time-out, corresponding to a kind of virtual complexity.

There have been moves to take an alternative perspective, to stand back and propose a solution to the problems of real-time programming in the form of a *global description* tool. First suggested by Mendelbaum and Madaule (1976), the point of departure is taken as the separation of global events and actions away from the description of each module, which may then be regarded as an autonomous primitive. Each event is associated with a sequence of actions, a reflex module to accommodate each exchange of information, similar to a B-activity.

Events, in this view, are of three types:

- an internal event, a fact of importance to be memorised for future sequencing operations;
- an expected event, a signal coming from the plant under control to notify the computer that some characteristic phenomenon has occurred;
- an unexpected event, arriving from an external source, but at an arbitrary time, such as an alarm.

Equally, these events are analogous to the following simulation complexity categories: temporal, configurational and interrupt respectively. This work is being followed up in the design of the language GAELIC (le Calvez *et al.*, 1977).

### 7.5.5 Summary

Although the fields of operating systems, multi-tasking, real-time systems and discrete-event simulation can be seen as varieties of communicating sequential processes, whether any further insights will emerge from considering such systems in the same terms, i.e. a CSP calculus (Hoare, 1985), remains to be seen. Although the different systems do contain many common concepts, such as events without duration, the motivations of their users remain distinct.

Consider for instance the famous example of the Dining Philosophers, described in more detail in Chapter 8. A student of operating systems is concerned with the possibility of starvation which the example can demonstrate, whereas for the simulationist the same model enables a comparison of wait-until facilities (Birtwistle, 1979).

The aim of concurrent programming is to encapsulate and separate individual concurrent processes to prevent unwarranted interference, whereas in simulation we are also concerned about safety, but we need a more intimate form of collaboration, at least to the extent of reading the attributes of another process instance. There is thus a basic disparity of interest, depending on the eventual application.

We find that rather than using the interrupt as a perfectly healthy control structure, many language designers have shied away, preferring to regard its occurrence as a danger signal. There is too much concern with the interrupt as a last-ditch attempt to restore a program or system from failure, but to *model* such situations requires a disciplined transferral from one set of interactions to another, without the stigma of an error condition.

Of the various kinds of complexity that have been investigated, some are already familiar from their implementation in simulation languages, and some may be novel. Besides inspiring a new look at the basic structures, this classification of complexity may prove useful in the following areas:

- to outline the features currently poorly supported by simulation languages;
- to enable a complexity comparison of simulation languages;
- to elucidate a comparison of simulation methodologies;
- to contribute to the discussion on canonical structures on concurrent programming;
- to provide a structure for implementing simulations on parallel hardware.

It is hoped that these themes will be followed up in future work. This charaterisation of complexity is currently being used in the design of the SIMIAN simulation environment, introduced in Chapter 8.

# 8

# The SIMIAN project

Throughout the book we have frequently encountered various elements of simulation language, more or less well defined, in both graphical and textual form. In this chapter we endeavour to bring these diffuse ideas together, under the umbrella of the SIMIAN simulation language project.

We start with the investigation of a basis for a semantic model of discrete-event simulation from which to derive the basic building blocks of the language. We find that a generalised concept of *wait until* pervades the kinds of phenomena we wish to describe, and this is used as the starting point for the language primitives. We then compare attempts to answer the vexed question of how to implement a generalised 'wait until'. Next, augmentations of the familiar Petri net are proposed to adapt it for simulation use. Thus enhanced, a Petri net will have three uses in discrete-event simulation:

- as an *intermediate model*, or *communicative model* (Overstreet and Nance, 1985) for an informal outline description of a simulation;
- in the form of a data structure to control the *activation* of the simulation;
- to provide a frame for the separable specification and automatic *collection of data* on the simulation run.

The implementation of a net-activated program forms the basis of the *engagement* (ENG) simulation strategy, which, together with a programming language and an interactive simulation environment, are the goals of the SIMIAN project. It should be emphasised, however, that this project is still in progress and many of the points made here are tentative and should be interpreted with caution.

## 8.1 UNIVERSALITY OF 'WAIT UNTIL'

One of the most widely discussed difficulties in simulation representation is the 'wait-until' problem, in which the start or outcome of an activity is determined by a particular *configuration* of the model. The time at which the event will take place is usually unknown when the preceding activity starts. The difficulty stems from the extensive nature of configurations, which may comprise a combination of several entities-in-state, perhaps involving functions of their attribute values. The problem is to ensure the event occurs as soon as the configuration arises. Birtwistle (1979) devotes a whole chapter to the problem.

The first paradigm of discrete-event simulation assures us that the basic type of event is *temporal*, for which the simulation executive knows exactly *when* to cause an event to be actioned, by the normal techniques of FES scheduling. The basic Petri net demonstrates, without any means of representing time, the dependence of transition firings on the occupancy of places, or, in simulation terms, the relationship between event occurrence and entities being in certain states. State-dependency is familiar from the simple customer–server interaction, where the need for a server in the idle state leads to the formation of a queue of customers. The fundamental division between time- and state-events has already been emphasised by Nance (1980).

Besides mere occupancy of places by tokens, we have seen how a Petri net can be enhanced to express the truth-value of a predicate-logic formula based on place occupancy, or *configuration*. Such values of model-configurations correspond to Reisig's *net predicates*, which are modelled in a diagram by *facts* (Reisig, 1985).

The two fundamental ways of defining the occurrence of a future event can be seen as an analogy of the ways in which the extent of an iteration may be specified in an algorithmic language, such as Algol 68. If we are dealing with a loop for which every iteration requires a new index value, and it is to be executed for a predetermined number of times, then we may use the **for** control structure.

**for** *i* **from** *start* **to** *finish* **do** *body* **od**

This statement will execute *body* for values of the index variable *i* as it is incremented from *start* to *finish* in unit steps. Given these values we can calculate exactly the number of times *body* will be executed. But sometimes this predetermined form of loop is inconvenient, and we wish to continue the iteration until a specific situation arises. Thus

**while** *cond* **do** *body* **od**

will continue iterating as long as *cond*, a Boolean expression which is re-evaluated at the start of each loop, has the value **true**. We do not need to know, nor can we usually tell, how many times the iteration will be performed before the Boolean expression becomes **false**. In order for the loop to ever terminate, it is of course essential that the value of *cond* is influenced, however indirectly, by the execution of *body*.

In the **for** loop, one particular variable, the loop index *i*, is extended over a given range; its reaching the limit of that range signifies the end of the loop. For the **while** case, between successive iterations, a particular condition (which may involve several variables) is evaluated and iteration is continued or terminated on the basis of the result of the evaluation.

In simulation terms, the 'future event' corresponding to termination of the **for**-loop is given beforehand in terms of a future event entered into the FES to occur at an exact time in the future. (This analogy is particularly strong for Algol 68 since the **for**-loop parameter values are unchangeable once the loop is entered.) But where loop termination requires a certain condition on the values of program variables to hold, and a **while**-loop is more appropriate, the analogy in simulation is some form of *wait-until* (*configuration*) statement. Notice that both **while** and *wait until* are predicated on a condition which is called by name.

The problem of handling configurations is basically one of arranging things so that the emergence of the correct combination of states can *activate* an awaiting entity. Approaches to tackling this task are complicated by the simulation strategy being employed. We have seen the facility which the PI strategy gives us in the resource assumption, obviating the need for a lot of standard definitions and activations. But the PI strategy, as it is commonly interpreted, has a difficulty in handling an all-or-nothing multi-resource acquisition. Birtwistle (1979) treats this in his chapter on wait-untils in DEMOS.

On the other hand, when using the AS strategy, a multi-resource acquisition is really no more difficult than one for a single entity, as any kind of predicate on the configuration is written as a C-activity test-head. Here lies the secret of AS's continuing success in the face of PI's elegance, efficiency and modularity. Since a Boolean expression comprising *any* program variables can be installed as the test-head of a C-activity, no special facilities are required to get the effect of a wait-until statement. But AS languages remain at too low a level in their lack of recognition given to simulation constructs, AS programs being essentially just sets of procedures. We should seek an approach which combines the best properties of both strategies.

One way of automating activations is through the provision of high-level constructs. While for commonly occurring predicates, such as 'resource is in idle state', this can be carried out by specific language components, such as the familiar ACQUIRE of DEMOS, there will always be some type of predicate which is just outside the reach of the language, no matter how richly endowed with individual constructs.

A sounder approach for the design of a language would be to generalise the phenomena of interest, and pick out the common concepts on which to form a basis for the language. Here we demonstrate the universality of the wait-until concept which can explicitly recognise model predicates. Once this is done, convenient language constructs can be synthesised from a fundamental 'wait_until (predicate)' form which can provide a basis for a built-in language semantics.

Taking the value of time as a component of the model state,

$$wait\ until\ (configuration)$$

can readily be seen as containing *temporal* complexity, i.e.

*wait until* (*temporal event*)

as a special case. Can we extend a generalisation of wait until to cover all seven types of complexity?

The resource assumption involves a special kind of dependency, in that the particular conditional state is one of availability. We adopt the convention of calling the state 'idle', to denote the available state of a busy–idle cycle. The customer entity waits for a server to attain its idle state:

*wait until* (*resource* IN *idle state*)

On the occurrence of the configuration, the two entities, customer and server, merge their identities for the subsequent engagement.

A similar wait-until situation prevails with *select* complexity; in this case the decision as to which of the entities will merge together is dependent on some criterion:

*wait until* (*attribute match*)

Until the occurrence of a suitable match, the entity waits. Thus we have been able to interpret the first four linear simulation complexities in terms of a general form expressed as

*wait until* (*predicate*)

where *predicate* is some state of the model, a model property, which must hold before the wait can terminate. We should re-stress here that by 'waiting' we refer to the entity's being attached to a particular engagement, irrespective of whether or not the engagement represents real waiting in a state of idleness or performing productive work.

Continuing with this interpretation, the three non-linear complexities can be seen in a similar light. While the divergent branching topology of *route* complexity does not necessarily imply any passage of time, two separate aspects of the same entity may wait for one another before merging. Many languages which incorporate multi-tasking (e.g. PL/I) include facilities for one task to wait for the completion of another. *Virtual* complexity is a wait for the first of a set of occurrences of virtual brother entities. The often-quoted example of the impatient lift-user, where the waiting is subject to two virtual outcomes, was used as a test for the wait-until constructs proposed by Vaucher (1973).

If we see *temporal* complexity as expressing the simplest, unelaborated form of outcome, then at the other end of the scale we have the *interrupt* which can be seen as a wait until exercised globally over the whole life-cycle, or even over the model as a whole, depending on its severity. On occurrence the interrupt can cause the current states of entities to be mapped into another model.

The wait-until characterisation is thus complete in the sense that, apart from divergence, all forms of simulation complexity can be seen as special cases. Of course, this is not to say that *all* representational problems of simulation have been solved, as we shall discuss later. But it does mean that we can attempt to unify discrete-event phenomena in a coherent form. For the meantime, we will proceed to investigate various suggested implementations of *wait until* as a basis for defining the semantics of simulation programming languages.

## 8.2   COMPARISON OF WAIT-UNTIL IMPLEMENTATIONS

The controversy over whether a language should include a wait-until statement, and if so, what form it should take, has been an issue since the first PI-strategy languages emerged in the early 1960s. While temporal events could be scheduled and the appropriate reactivations made at the due time by the simulation executive, configurationally dependent events are not so easy to automate. This is because temporal events are strung out in a one-dimensional continuum of time, and are therefore relatively easy to detect. But the set of configurations may have one dimension for each state on which the conditions are predicated.

Not only must the critical configurations be detected, but the awaiting entities must be reactivated. As we saw with the ES strategy, expressing reactivations at the point where the condition is detected can lead to a poorly structured program, which can be difficult to get right and to maintain and modify. It is easy enough to forget to reactivate an entity, resulting in a program which appears to 'work' but does not truly reflect the dynamic of the real system.

Pursuing the wait-until question has been carried out mainly in PI-strategy terms, usually in the context of Simula. The original wait-until construct of an early version of Simula:

> *pause (Boolean expression)*;

was quietly dropped (Nygaard and Dahl, 1978), on the grounds of its inefficiency, and replaced by *passivate, activate* and *hold*. This decision, avoiding interrogative sequencing, was based on the conviction that the language must give full sequencing control to the user. However, the case for interrogative sequencing (Blunden, 1968) still has certain advantages.

A wait-until statement is easier to use and quite accurately describes itself (Nygaard and Dahl, 1978). It is more self-contained, leading to a better program structure. If implemented properly, it can prevent the user from making the error of forgetting to reactivate the process at the proper time. This kind of omission is particularly serious, as the observable consequences are negative: things that should have happened do not. Despite these recommendations, it was decided that Simula should concern itself only with *imperative* sequencing, for two reasons.

1. The idea of waiting until a given condition becomes true is not well defined for quasi-parallel processes. The stage of a program at which a condition becomes true may be influenced by implementational factors.
2. There is no upper limit to the time required to execute a wait-until statement.

As experience with Simula grew, and comparisons were made with other simulation languages, such as GPSS and SIMSCRIPT, the conclusion was drawn that the facilities provided by Simula needed a further level of sophistication before they would approach the convenience of other languages. For instance, the ENTER command of GPSS contains many features which must be individually specified in detail in the Simula equivalent. This is clear if we compare the concise GPSS program in Fig. 6.10 with the Simula version in Fig. 6.12. While Simula provides the essential requirements for simulation, such as coroutine processing, future-event scheduling, etc., it was generally felt they were not sufficiently well packaged for general use.

Early experience with Simula had shown how easy it was to forget to include reactivation of waiting process instances when they should be triggered (Dahl, p. 362 of Buxton, 1968). One approach to cure this problem is to try to emulate the structures of GPSS in Simula (Vaucher, 1971; Houle and Franta, 1975). The most difficult concepts to carry over from GPSS to Simula were features for *route* complexity, such as GPSS's TRANSFER statement, and the configuration-checking performed by the TEST statement, which requires the so-called 'generalised wait-until construct' (Vaucher, 1973).

On the other hand, for AS languages, wait-until specification was no special problem, being just another C-activity combining a conditional test-head with an action body. Dahl (1968) summed up the separation of concerns with his division of simulation languages into interrogative and imperative languages. Meanwhile, AS-language designers were more concerned to improve the efficiency of the C-phase, i.e. to develop an automatic method for restricting the breadth of the C-phase scan to only those C-activities which could possibly be affected by the actions of the just-completed B-activity.

At the same time, the question of *contention* between C-activities was raised. Contention may arise if, for instance, a resource released in the B-phase is required by more than one C-activity. Which C-activity should acquire the resource? Various solutions were suggested, such as changing the scanning order of the C-activities, giving the C-activities priorities (as in ECSL, Clementson, 1977), or randomising the scanning order, but the general consensus was that it would be too confusing for the user to specify, and only a few attempts at resolving this problem ever saw the light of day. We will consider the contention problem in connection with the engagement strategy later in this chapter.

There are some wait-until conditions which cannot be monitored by facts, because they involve time changing in a continuous fashion, and thus fall outside of the first paradigm of discrete-event simulation. Such conditions, when they are of sufficient significance with respect to other features of the simulation, properly require a combined discrete-event and continuous approach to simulation. In this case, the wait-until construct can serve as an interface between the two models, allowing engagements in the discrete-event part to be conditioned on the attainment of thresholds in the continuous part, and coefficients in the continuous part to be effected by engagement formation and dissolution in the discrete-event part.

In the restricted case where the flows can be assumed to be linear functions of time, a three-phase AS executive can be enhanced to deal with events arising from continuous flow as if they were 'ordinary' future events (Evans, 1981, 1983). The same kind of features are also instrumental in bringing about event occurrence based on the accumulation of a specific amount of time spent by an entity in an activity (or set of activities). Such events are especially necessary in the simulation of breakdown based on the expiry of an entity's working life.

The early PI language SOL (Knuth and McNeley, 1964a, b) has a wait-until command

**wait until** (*condition*)

which stores transactions according to the kind of condition awaited. For instance, if the condition is the attainment of a certain value of time, then the transaction is queued on the FES, in its appropriate place. The transaction suspended by the statement

**wait until** $(A=0)$

is stored in a set associated with $A$, and is only interrogated when the value of $A$ changes, in a simple activity-scan (Laski, 1968). But the user is warned not to invoke, overtly or as a side-effect, any random expression within the condition definition. This construct resembles the *pause* statement of Simula which was dropped.

Vaucher (1973) proposed a generalised wait-until simulation language construct for Simula of the form

*wait until* (*condition*)

where *condition* is a Boolean expression of arbitrary complexity. The semantics would be that a process instance executing such a statement would automatically become passivated until the condition became **true**. The implementation required the automatic setting up of a list to contain process instances awaiting the condition, and a monitoring procedure to oversee reactivation. A global Boolean variable *action* used for communication between blocked process instances and the monitor was also required.

The constructs can be defined using the *simset* level of the Simula

CLASS hierarchy (Birtwistle *et al.*, 1973), which enables two-way linear lists to be set up, prefixing the *simulation* CLASS, as in Fig. 8.1.

Notice that the Boolean expression *b* is called by name, which means that every time it is invoked it takes a fresh copy of its component variable values, so its result reflects the state of the model at invocation time. If *b* evaluates to TRUE, the invoking process instance is allowed to proceed; otherwise it is passivated and stored in the *waitq*. When the monitor eventually reactivates the process instance from the *waitq*, the condition *b* is re-tested, and the process instance returned to the *waitq* if it is still not TRUE. Otherwise the *waitq* is quit. The post–initial testing of *b* is controlled by the monitor, which itself is nudged into action by a call from *wait until*. When a process instance finally does quit the *waitq*, then the monitor is notified by the *action* variable taking on a TRUE value.

It is the monitor's responsibility to ensure that passivated process instances are reactivated at the 'right' time, i.e. as soon as the *b* expression could be TRUE. It is clearly inconvenient to repeatedly attempt to remove the process instance from the queue at times when *b* could never be TRUE.

The declaration of the *wait monitor* is given in Fig. 8.2. Clearly a process instance passivated by a *wait until* must remain passivated at least until the system state has altered, because only such a change can influence *b* and its referents. Any process instance booked on *waitq* must be activated to re-check *b* after every event occurrence. The variable *nextev* refers to the next event-notice in the FES. When a process instance is released from the *waitq*, this happening is transmitted to the *wait monitor* (by *action* being TRUE) and the monitor continues to select further process instances from the *waitq*.

There are several criticisms that could be levelled at this proposal. It seems that all configurationally determined events must share the same *waitq*. This would not be too inconvenient, were it not for the fact that the monitor will attempt to release process instances from the *waitq* in FIFO order, which may induce an artificial ordering on activations, since the standard *simset* queueing routines are used. A FIFO relationship between instances in the case where there are two *wait untils* operating at the same time might be unfair. Having a separate *waitq* for each condition would also be more efficient in that a single state-change should not necessarily imply a test of all waiting instances. Franta (1977) notes that the interests of efficiency are better served by having one *waitq* for each wait-until condition. Further inefficiency is caused by the large proportion of tests being unsuccessful.

Vaucher (1973) correctly notes the unfortunate consequences which might result if *wait untils* were used in place of the standard FE-scheduling routines. Consider the following statements:

(a)   *wait until* ( *time* = 15 );
(b)   *wait until* ( *time* > 15 );

given that the current time is 10 and the next event on the FES is at time 20. Then at time 10, neither statement will be activated, whereas at time 20 (the

REF (*head*) *waitq*;
REF (*wait monitor*) *monitor*;
*waitq*:– NEW *head*;
*monitor*:– NEW *wait monitor*;

PROCEDURE *wait until* (*b*);
NAME *b*; BOOLEAN *b*;
BEGIN
          IF *b* THEN GOTO *exit*;
          *into* (*waitq*);
          IF *monitor.idle* THEN ACTIVATE *monitor* AFTER *nextev*;
*loop*:   *passivate*;
          IF NOT *b* THEN GOTO *loop*;
          *out*;          COMMENT i.e. quit the *waitq*;
          *action*: =TRUE;
*exit*:
     END *wait until*;

Fig. 8.1 — The *wait-until* procedure of Vaucher (1973).

*process* CLASS *wait monitor*;
BEGIN
          REF (*process*) *pt*;
*start*:  IF *waitq.empty* THEN *passivate*;
          *action* := FALSE;
          *pt* :– *waitq.first*;
*loop*:   IF *pt* == NONE THEN GOTO *wait*;
          ACTIVATE *pt*;
          IF *action* THEN GOTO *start*;
          *pt* :– *pt.suc*;
          GOTO *loop*;
*wait*:   REACTIVATE THIS *wait monitor* AFTER *nextev*;
          GOTO *start*;
     END *wait monitor*;

Fig. 8.2 — The *wait monitor* of Vaucher (1973).

next event), the process instance using (a) will not have been activated. but (b) will be reactivated at time 20. Neither of these results suits what is presumably intended by the above statements. It is important to remember that *wait until* cannot bring extra temporal events into existence; the set of significant discrete events is still controlled by the retrievals from the FES.

In the Aspol simulation language (MacDougall and McAlpine, 1973; MacDougall, 1975), the primary concepts are the process and the event. Once a response is initiated, it exists on one of four states: *execute, ready, hold* or *wait*, until it terminates. To suspend a process for a known period of time, we have the *hold* statement, familiar from Simula. Aspol is different from other PI languages in that it recognizes an event structure. Events have identifiers which take on the values *occurred* or *not occurred*. Events provide the channels through which processes communicate and coordinate their actions. They are defined by declarations

**event** *e1, e2*;

and can be assigned values. A call of *set* (*e*) means that the value of event *e* becomes *occurred*. Processes may suspend until an event *e* occurs by either *wait* (*e*) or *queue* (*e*), an action which is called *event selection*, and the suspended processes are called *event selectors*. If the event already has the value *occurred*, the process continues. Otherwise, when the event in question eventually occurs, all *waiting* selectors and one *queueing* selector (the one with highest FIFO priority) are put into the *ready* state. Houle and Franta (1975) give a Simula version of Aspol's event structure, with the extra capability of priority queueing.

A simple PI simulation language, SIMON 75, containing automatic wait-until scheduling statements was described by Hills and Birtwistle (1975). The language views an entity processing through an activity as a series of four phases:

- WAIT UNTIL the resources are available;
- COOPT the resources to the entity;
- HOLD for the direction of the activity;
- RETURN the resources.

Within this framework, SIMON 75 aims to provide routines for specifying a wide range of problems, including resource types, data collection and report routines, and a fail-safe mechanism for carrying out activations. An entity can also offer itself as a resource by entering a WAIT (Q) statement, in which case it can be COOPTed, or another entity may FIND it.

In order to implement the four phases of SIMON 75 and the fail-safe activation mechanism, each time-beat incorporates a scan of possible reawakenings, similar to an AS C-phase. After resources have been released, the passive entities which are in the WAIT UNTIL phase are reactivated, in the order in which they are declared, so that their conditions for continuation are re-checked. This is the so-called *Snoopy* approach for reactivating entities, being more efficient than the 'pure' AS strategy which tests all C-activities during each time-beat. The language thus represents a compromise between AS and PI strategies.

DEMOS (Birtwistle, 1979) has four kinds of synchronisation statements, including the pair WAITUNTIL/SIGNAL, in which a Boolean expression can be made the condition for entity advancement. Waiting entities are queued on a condition queue, one for each condition. Also considered are more complicated selection rules where an entity may wait to see what other entities become available before deciding what to do, and which entities to do it with.

Birtwistle also gives an interesting comparison of two competing approaches for the performance of conditional restart. SIMON 75's *Snoopy* is compared with DEMOS's *signal* statement in which the responsibility for reactivating passive entities is left entirely to the programmer. In the case of the DEMOS language, the designer has simply washed his hands of the responsibility for reactivation and leaves it up to the simulationist to be on the lookout for situations where it might be possible that the condition could become true, and include some explicit testing code at those places. This is the philosophy of its host language, Simula.

Further support for the view that a wait-until concept underlies simulation constructs comes from the implementation of the synchronisation constructs of DEMOS. The four DEMOS synchronisation pairs, ACQUIRE/RELEASE, TAKE/GIVE, COOPT/SCHEDULE, WAIT UNTIL/SIGNAL, can be expressed in a uniform way in terms of a more primitive pair GET/PUT (Birtwistle, 1979, p. 159), informally described as follows:

> PROCEDURE GET (Q, AVAIL, SEIZE);
> NAME AVAIL; REF (QUEUE) Q; BOOLEAN AVAIL;
> statement SEIZE;

and

> PROCEDURE PUT (Q, RETURN);
> REF (QUEUE) Q; statement RETURN;

where 'statement' is an imaginary Simula type. These primitives are concerned with representing an entity entering into and exiting from a queue, paying attention to the case where the new entrant happens also to be the first in the queue. Since all waiting sets in DEMOS are implemented as some form of queue, the GET/PUT pair can generalise all DEMOS synchronisations, so that synchronisation pairs can be defined as special cases of GET/PUT. But of these, the WAITUNTIL/SIGNAL pair requires the least elaboration, demonstrating that, for the DEMOS case, wait until is the most primitive synchronisation.

Another proposal for efficient implementation of a wait-until construct is advanced and discussed by Lapalme and Vaucher (1981), who use data-flow analysis to determine interactions between processes to reduce the overhead in wait-until implementation. They consider the case where an event occurrence allows several process instances to be released from waiting for the same condition simultaneously. If they are released sequentially, then it could happen that released instances attended to later could find that the conditions for release no longer hold. In the absence of any standard means to specify priorities between such process instances, the authors prefer a random choice of events whose condition is true, although it is clearly better to allow the programmer to specify the rules rather than apply an *a priori* rule.

The authors leave undefined the priorities between different conditions, and attend to the process instances awaiting the same condition in FIFO order. This rule is only applied in default of more detailed instructions being available from the simulationist. In defence of this rather arbitrary rule, it is pointed out that many situations, such as real-time and distributed programming where synchronisation by condition is required, are similarly arbitrary.

To achieve this policy, it is sufficient to test each condition after every event occurrence, as the three-phase AS executive does. But it is necessary to test only a subset of the conditions, namely those effected by the nature of the state-changes performed at event occurrence. The analysis of the effect of component statements is undertaken to reduce the cost of the condition checking to an acceptable level.

For implementation, the conditions are classified by different schemas, before transforming the source program by a pre-processor into Simula code. Any conditional-wait program can be transformed into an imperative form, provided that, at all points of interaction, reactivation of waiting processes which have the chance of moving by the effect of previous statements is included. In this way we can be assured that the necessary reactivation logic is not omitted.

The approach of Lapalme and Vaucher is to categorise the conditions according to the variables which they contain. There are three such categories of conditions:

- containing global variables only;
- containing local variables, but no function calls; and
- containing function calls.

Each category is implemented in a separate way. The conditions can be further broken down according to criteria which indicate the kind of change to the variable which would tend to make the condition true. For instance, if the condition is

$$a > expression$$

then only an *increase* in the value of $a$ would possibly make the condition true. For

$$b < expression$$

only a *decrease* in $b$ would make the condition true. In more complex conditions, e.g.

$$a-b > 0 \text{ and } c \neq 1$$

the condition would be subject to three criteria of change:

- *a*: increase;
- *b*: decrease;
- c: any change.

Each variable appearing in a condition must thus have the appropriate monitoring placed around any write-access to it, and if the change is of the correct sense, a reactivation of the processses awaiting the condition is re-attempted. Notice that if in the above expression **and** were replaced by **or** then exactly the same categorisation would be made. Categorisations for other logical operators are made accordingly.

The amendments to the Simula program making use of the wait-until construct are carried out by a pre-processor, giving what the authors call an *optimised wait-until implementation*. In tests, it was shown that consider-able improvements in run-time could be obtained, justifying the whole scheme. The authors recommend the ideas should be transferred to other areas involving parallel processes.

Laurini (1981) suggests a formalism for predicate-driven simulation including a feature with a quite readable form for specifying configurational events:

**when** *predicate* **activate** *process*

directly linking the process with the condition for which it waits. A Simula implementation of these ideas has been made, which has been used for the simulation of urban processes (Laurini, 1979), but the implementation is not described in detail.

A large amount of confusion still surrounds the discussion of wait until. No implementation of a general wait-until construct with automatic reacti-vation can be said to be widely accepted, despite its attractions. Moreover many approaches require language pre-processors, which can be unwieldly, denying the simulationist the flexibility required by a 'what-if' investigation. The most thorough-going approach, that of Lapalme and Vaucher (1981), treats the simulation as just a Simula program, structured at the level of program identifiers. Perhaps an analysis of a simulation at the level of its Petri net would be better

A comparison of strategies leads us to consider an important question for simulation language design: how best to combine the conciseness and modularity of PI with the universality of AS? The question necessarily involves the handling of wait untils, and naturally leads on to the specifica-tion of how to resolve contentions. Can a language be designed out of an effective marriage of PI and AS? We hope to shed some light on this question later in this chapter in the proposal for an engagement strategy.

## 8.3   AUGMENTED PETRI NET

### 8.3.1   Homomorphic simulation
As we have already mentioned, the usefulness of models lies in their ability

to represent the essentials of a simulation — a good model contains only the relevant details and discards superficialities. What is relevant is best described with regard to the intended purpose of the model, and a model may be judged by the aptness of the component aspects selected for inclusion by the modeller. If every possible aspect were included, the model would be nothing but an exact copy of the intended reality. Such closeness is not usually desirable, as it defeats the intention of modelling. In mathematical language, we can say that a model bears a *homomorphic* relationship to its referent, insofar as it preserves essential qualities, while rejecting others.

In addition to this rather obvious property of all models, the word 'homomorph' is also used in a restricted sense to refer to a device to improve the efficiency of an AS simulation. Since, for any time-beat, only a single B-activity is called, there will usually be only a subset of the entities whose states will change as a direct consequence of the activation of the B-activity. Also, for any C-activity, the test-head will refer to only a subset of the entity states. Thus we can expedite the C-phase of the executive by testing only those C-activities which could be affected by the B-activity activated in the B-phase.

In order to implement this scheme, a mapping from each B-activity to the whole set of C-activities, usually in the form of a bit-pattern (Tocher, 1969), is necessary. The problem is how to set up the mappings. If the simulationist performs the task, then he is being required to specify what is essentially the same information twice. Any incompatibility would be difficult to detect, yet the simulation could very likely contain errors caused by activation omission. Alternatively, and more securely, the compiler could extract the information from the original definition of the simulation.

Birtwistle (1979) remarks that such a facility would make AS languages more competitive with PI languages in run-time efficiency. However, the automatic production of this kind of mapping would present a considerable task, especially as the contents of C-activity test-heads are not confined to entity attributes, but may contain references to *any* program variable. It would also entail following up all the possible side-effects of procedure calls.

Basically, what we propose in the engagement strategy is to make use of the Petri net of the simulation, being a homomorphic mapping between successive engagements of an entity's life span, to perform as the activator of the successive events (C-activities) of the entities' processes. The arcs are used to activate only those transitions which could possibly be enabled by the presence of entities in states (or, tokens in places). In this section we describe the necessary augmentations that we apply to a Petri net to enable it to represent a discrete-event system more adequately.

### 8.3.2   Basic types of Petri net

To recap, the basic Petri net (see Fig. 6.3) consists of just four components: transitions, places, arcs and tokens. Arcs, which are directed, connect a place to a transition, or a transition to a place, giving, for any successive sequence of arcs, a strict alternation of place and transition. Tokens can reside on places, and can be subject to movement to other places by the operation of a *firing rule* for a transition. For any transition, there is a set of input places, consisting of all the places which are connected to the transition by an arc directed towards the transition. Similarly, there is a set of output places connected by an arc directed away from the transition. The firing rule causes tokens to be removed from input places and placed in output places. The rule operates only when each of the input places contains a token.

In an interesting discussion, Reisig (1985) mentions four interpretations of a Petri net. In the condition/event (C/E) net, places may be marked, by having a resident token, or unmarked, whereas for the place/transition (P/T) net, the *number* of tokens resident at a place is significant. This simple extension means that P/T nets can be subject to a firing rule, and can be used to study system phenomena such as liveness and deadlock, familiar from computer operating systems.

Extending Petri nets to a higher level by allowing tokens to be regarded as individuals, we get a predicate/event (P/E) net in which relations between individuals can be formulated in predicate logic by the use of facts (Genrich and Lautenbach, 1981). By making the same extension from C/E to P/T, i.e. by allowing tokens to represent a number of individuals of the same kind, and applying it to the P/E net, we get the *relation net* which is the richest form of net discussed by Reisig (1985).

### 8.3.3 Petri net for simulation

Genrich and Lautenbach (1981) introduced the idea of system modelling with high-level Petri nets, using tokens to represent individuals. We start with the *relation-net* abstraction of the Petri net in which all tokens are considered as individuals and any number of any kind of token can occupy a place. Clearly, the relation net is very similar to, although more formally defined than, the activity-cycle diagram, as described in Chapter 6. For a complete description of a discrete-event system, the main thing which is lacking is the concept of a *time-delay*. The Petri net is primarily concerned with 'state-dependency' of happenings, in that the firing depends solely on the occupancy of places by tokens, and in its basic form has no means to represent the passage of time. As soon as the occupancy conditions are fulfilled, the transition fires.

The way in which we have chosen to represent temporal dependency is in the form of a star symbol connected by a directed arc to a transition (Fig. 6.5). The star can be thought of as an extra place which receives a token instantaneously, when and only when the corresponding event should occur. Also, the token vanishes on reaching the transition. The occurrence of an event is represented by the firing of the transition, and the change of state of an entity is brought about by the removal of tokens from input places and placing tokens in output places.

The relationship between an engagement and its representation in net terms is given in Fig. 8.3. In addition, tokens may be created or destroyed at a transition. Furthermore, a transition firing can cause the merging of tokens to form composite tokens, or the splitting up of composites into smaller composites or individual tokens. In short, anything which can happen to a

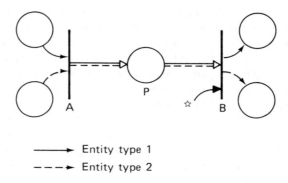

Fig. 8.3 — A typical engagement in net form: at transition A, an engagement
consisting of a collaboration between entities of type 1 and 2. The place P gives the
name of the activity of the engagement and transition B is the outcome.

token happens when a transition fires; a token remains unchanged while it
resides in a place.

Entity types may be distinguished by being represented by tokens of
different colours, and the arcs permitted to tokens take on the appropriate
colour. For engagements of collaborating entities, this means that some
tokens and arcs are multi-coloured to represent the mixture of entities.
Other descriptions of coloured Petri nets have been made, involving
coloured transitions and places (Peterson, 1980; Jensen, 1981), with the
intention of shortening system description. In our description, token colour
denotes entity type, but each entity may be endowed with further attributes
which can be used in allocation decisions to influence whether or not a
transition will fire. The attribute information is not necessarily represented
in the net.

These identifications between a discrete-event system and the elements
of a Petri net can be summed up in a table, Fig. 8.4.

| transition | — | event |
| transition firing | — | event occurrence |
| place | — | activity, or state |
| token | — | entity |
| coloured token | — | entity of a certain type |
| arc | — | sequencer of entity flow |
| coloured arc | — | route followed by entity type |

Fig. 8.4 — Correspondence between net elements and a discrete-event system.

The totality of tokens in places represents the instantaneous *model
configuration*. To model events which react to configurations, we can
indicate the dependency by means of a *fact*, as in Figs 7.1 and 7.2. A fact is
connected to places and transitions by *control arcs*, expressing that a

predicate holding over the places may affect the firing of the transitions. No flow of tokens along control arcs is suggested. Similar arcs, called *information arcs*, are required to indicate dependencies between transitions and tokens-in-places in the representation of *virtual* complexity (see Fig. 7.5).

It should be stressed here that what is represented in a net may not correspond in detail to actual entities. We do not aim to model at the level of attribute values, but to pursue the activities of entity types. This means that sometimes we need to abstract the net away from some details which are necessary in the actual simulation itself. Consequently we must look upon the net as a statement of what is *possible* for an entity type, without implying that *all* entities of that type will necessarily follow any particular route on every occasion, as such decisions are often based upon attribute values.

As an example of an augmented net, let us consider the description of a ship entering a harbour (Fig. 8.5). A ship wishing to enter a harbour must have a berth allocated to it, a tug to negotiate the entrance, and the tide should be high enough to allow its passage. These conditions granted, the ship enters the harbour, after which it releases the tug and undergoes unloading. When unloading is complete, a tug and high-tide are again required for exiting. There are four entity types: ship, berth, tug and tide, represented by different token 'colours', as given in the key. The *high* place for the tide is monitored by a fact, which enables transitions ending *wait-in* and *wait-out* states. Both tug and berth entities are resources, comprising different kinds of busy–idle cycles.

Peterson (1977) mentions the hierarchical modelling property of Petri nets, in which an entire net may be replaced by a single place or transition (or vice versa) for modelling at a more abstract (or more refined) level. This is a useful property, allowing phenomena which do not require explicit model ling to be subsumed under the umbrella of a single component. For our augmented nets, such abstraction may not be so readily applied, but we can conceive of subnets as being replaced black boxes with the terminal elements of the proper kind, so as to fit the alternation of place and transition.

The convention coined by our augmented Petri net is that the generation of an entity is associated with a transition, designating it thus as an instantaneous event, and the termination of an entity is into a place without an exit arc. The reason why we choose termination to be a place is that the experimenter often needs to keep account of how many entities have flowed through the system — this fact is easily ascertained from the augmented net by monitoring the termination place.

As we mentioned in Chapter 6, it took a long time for simulationists to recognise that their own version of asynchronous systems, which we call here discrete-event systems, had strong similarities with others, and moreover that the diagrammatic forms of representation for both were also similar. Törn (1981) was the first to make direct comparisons between the Petri net and the activity-cycle diagram. In Törn's paper, transitions of the net correspond to events or activities (presumably of the B- or C-type, although Törn later defines an activity as 'a number of closely related

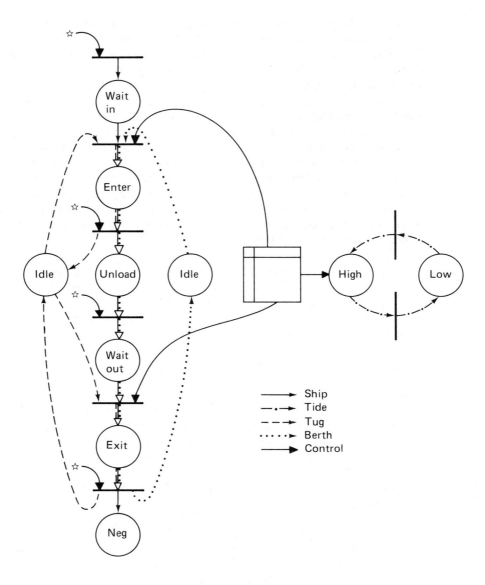

Fig. 8.5 — Petri net of harbour simulation.

events'), and places correspond to conditions necessary for the event or activity which follows to occur. Noting that time-delay is not readily represented in Petri nets, Törn goes on to define an extension of the Petri formalism called *simulation graphs*.

The basic Petri net is enhanced by having a time notation, by which each transition has an allotted time to complete the token movement associated with its firing. Naturally enough, the time specified can be a sample taken from a probability distribution. Further enhancements include multiple arcs, inhibitor arcs, test arcs and a queue as a special kind of place. In a subsequent paper, Törn (1985) adds interrupt arcs and coloured tokens, and describes a simulation-program generator, SIMNEX, based on these ideas. Somewhat similar ideas have been developed for representing real-time systems (le Calvez *et al.* 1977).

### 8.3.4 Homomorphic hierarchy

Overall, in the process of producing a simulation of a real system, we construct what amounts to a series of models, each of which bears a particular relationship to its predecessor. The structure of models is specified at different levels, and correspondingly there are different levels of shared structure or homomorphism (Zeigler, 1976). The total simulation activity, from apperception of reality to the architecture of implementation, which we hope to capture in the SIMIAN project, may thus be viewed as a homomorphic hierarchy, Fig. 8.6.

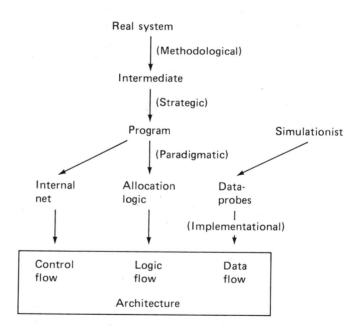

Fig. 8.6 — The homomorphic modelling hierarchy. The sequence of transformations performed in simulating a real system.

We start with the real system, which may be real to the extent that parts of it exist hypothetically in the mind of the simulationist. As a first step, the intermediate model is propounded, which serves to eliminate extraneous detail and concentrate attention on the main processes involved in the model. As we have already emphasised, the intermediate model is best

represented as a diagram, perhaps informally annotated, which is a convenient form for discussion among non-technical members of the design
team. It may take the form of an activity-cycle diagram or a Petri net.

Once formulated, the intermediate model can be used to help in the
construction of the simulation program, written in a simulation programming language. This step requires the acceptance of a simulation strategy for
representing the system. Elements of the simulation language can be broken
down into more primitive forms, which can be used to define the semantics
of the program. In a later section of this chapter we will suggest a set of
suitable simulation primitives. Based on the second paradigm of simulation,
the simulation program itself can be subdivided into three components:
engagements, allocations and data-probes.

The sequence of engagements is represented by an internal net which,
although resembling a programming-language version of the intermediate
model, takes the form of a data structure to direct the activations of the
program. Catering for the particular tactics to be adopted in dealing with
questions of priority and contention between entities comprises the bulk of
implementational detail of the engagement strategy. Allocations, in which
we specify exactly how entities are chosen to engage with one another, are
described in program modules, and can be regarded as being attached to the
transitions of the internal net — they correspond to 'extra conditions'
imposed on the normal firing rule.

The third component, the data-probes, do not reflect any aspect of the
real-world system, but are measuring devices inserted at transitions by
which the simulationist may collect statistics to gauge the dynamic behaviour
of the system as the program is running. Inserting, repositioning and
adjusting data-probes greatly facilitate system investigation, so it is important that they should be implemented as components which are quite
separate from the rest of the simulation program. Of course, their presence
must not be allowed to influence the behaviour of the model.

Methodologically, manipulation of the data-probes belongs to the experimentation stage and is better not to be considered as part of the simulation
program. In practice most simulation languages deny the user this facility,
with the effect that every slight change to the data-probes requires recompilation. To avoid this, programmers are tempted to place data-probes at
every conceivable juncture, with the result that there is often an overpro-
duction of output. In approaching this problem, we must extend the limits of
the SIMIAN project to include the whole of the simulationist's programming environment.

There seems to be a close relationship between the three simulation
components and various programming perspectives which have been suggested by Nygaard (1986). This will be considered further in Chapter 9.


## 8.4   ENGAGEMENT STRATEGY

### 8.4.1   Engagements

Throughout the book we have often made reference, in different contexts, to the idea of *engagement*. In particular, the term 'engagement' has been used to denote four related concepts:

(i)  the state or activity of an entity;
(ii)  a data structure to indicate attainment of the state;
(iii)  the formation of a collaboration between entities;
(iv)  a stage in the development of a process.

In this section we bring these ideas together, and describe how they interact in the engagement (ENG) strategy.

The term originates from Tocher's AS-strategy language GSP (Bent, 1976), where it describes the current activity of a machine (i.e. a permanent entity, or resource), which can have either a permanent or a temporary engagement. If the duration of an activity is known, the machine is *committed* until the future event occurs which corresponds to the end of the activity. When this happens, the machine either returns 'idle', if the engagement was temporary, or returns 'engaged', and is subject to further re-engagement and commital. An uncommitted machine is regarded as being idle. In our engagement data structure, Tocher's idea of engagement in a state is retained, and combined with committal information, i.e. the set of outcomes, and the total set of entities collaborating in the activity. Thus together in one structure all the information required to effect *virtual* and *interrupt* complexities is available.

The sequence of engagements encountered by an entity is defined in the appropriate coloured-arc sequences of the Petri net. For any entity, all the time it spends in the system is accounted for by belonging to engagements. Changes of engagement take place only at transition firings, which are regarded as instantaneous happenings. In order to investigate the dynamic behaviour of a model, we therefore need to be concerned only with the changes of engagement occurring at transitions, to which we can attach data-probes. The real dynamic is thus mapped to the engagement sequence; state names can be given directly to the corresponding places on the Petri net. Other control flow in the program is contained in the data-probes and allocations, both of which operate instantaneously in terms of simulated time.

The allocations are a veritable part of the real system, but are concerned with the flow of *logic* rather than time. For instance, at every state-exchange, there may be a decision as to which particular entity of a collection belonging to the same state should next become engaged. Such decisions belong to the real system, and must therefore be represented, but since they are instantaneous they need not be monitored by data-probes. Any problems which occur with allocation, such as contention between entities for resources, can be reported on in whatever way the simulationist it thinks necessary, making use of environmental facilities. If it is necessary to model the time required to make a decision, then a decision-maker entity should partake in the decision-making activity.

While Wirth's concept of a program as

$$\text{program} = \text{algorithms} + \text{data structures}$$

is a suitable decomposition for algorithmic languages, for our purposes we state the second paradigm of discrete-event simulation:

$$\text{simulation} = \frac{\text{engagements}}{\text{data-probes}} + \text{allocations}$$

on which basis we pursue the development of SIMIAN. The engagement sequence is preserved in the sequence of places, while the allocation algorithms remain as program modules, invoked on attempting to fire transitions. Data-probes attached to transitions monitor the simulation by being triggered by the passage of tokens, during the firing of transitions.

For completeness, we should consider this development in terms of a total simulation programming environment, incorporating the system description in SIMIAN alongside the separately specified experimental control of the simulation run, which manipulates the data-probes and supervises the output of reports and statistics.

## 8.4.2   Engagement strategy

The main concern of any strategy is with the parallelism of a real system and its sequentialisation in terms of a program. From the viewpoint of a Petri net, we may have many transitions simultaneously able to fire, but in which order should we attend to them? For our augmented nets, this question is partly answered by the star symbol which applies an ordering between temporal transitions based on event retrievals from the FES; but still there remains contention. One of the aims of the strategy is to give expression to such non-determinism and enable the simulationist either to make an on-the-spot choice, or to propose an heuristic.

In the ENG strategy we combine two strategies: in the description of the real system we use a PI strategy, taking advantage of conciseness of expression and the state-sequencing information provided in the entity process. During compilation, the sequence information is incorporated in the Petri net, and activation modules (corresponding to B-activities and C-activity bodies) are composed, to be invoked by transition firings. The implementational strategy resembles a three-phase AS executive, with a selective C-scan, with a contention check preceding it, and automatic checking of wait-until predicates. The ENG strategy can be thought of as being composed of two scans:

(i)  over time (i.e. to get the next temporal event);
(ii) over configurations.

The first paradigm assists us in these scans, since it assures us that we can assume future events will contain all significant times and that configurations can only change, and therefore need only be checked whenever tokens move in response to some temporal event occurrence. State-changes are thus initiated by an event occurrence (or a transition firing, in net terms) and the effects are transmitted, following the arcs of the net, to neighbouring places. While the FES is a collection of temporal activations, the Petri net is a storehouse of configurational possibilities. The engagement strategy uses both structures in tandem. Each scan in turn consists of three phases:

(a)   selection of candidate events;
(b)   elimination of contention between selected events;
(c)   perform the actions associated with the event.

For the time-scan, candidate events are simply those future events with equal earliest occurrence time and should not cause too much problem if a real-number time-base is used, as we discussed in Chapter 3. Coincident configurational events are more difficult to deal with and are brought to the surface for consideration in the ENG strategy. Further events may be immediate consequences of performing the actions. After each action is performed, the configuration scan is applied until there are no more actions, when we perform the time-scan again, completing the cycle.

Few simulation-language designers have paid much attention to the problem of how to proceed when a configuration may cause one of a set of transitions to fire. Laski (1965), in a paper which has been largely ignored, attempts to grapple with this problem in CSL terms, but the concepts are hampered by being expressed as a combination of ES and AS (time-cell) strategies. As the author admits, the suggestions were never implemented in total owing to the unlikelihood that any single simulation would contain all of the complexities catered for, so the resulting system would inevitably be wasteful. A later paper by Parnas (1969) bases itself on a PI strategy and reports difficulties with its basic assumptions, as do Lauer and Shields (1978). Hopefully, by uncluttering the problem along the lines suggested by the second paradigm and use of the Petri net as an activator, we may throw some more light on the problem.

Strategies are rarely implemented in a completely pure form. Zeigler (1976) regards the three-phase AS strategy as a mixture of ES and the time-cell variant of AS. SIMSCRIPT II.5 offers the option of either ES or PI strategies. The ENG strategy is a compromise between PI and three-phase AS (Fig. 8.7). The controversial inclusion of *Snoopy* reactivation in SIMON 75 can be seen as another way of reconciling PI with the C-phase of AS.

While we have shown, in Chapter 6, that the PI strategy possesses certain advantages over others, particularly in the area of preserving the information presented by an activity-cycle diagram, there are also some pitfalls associated with the PI strategy. Despite its name, the PI strategy's claim to be concerned with *interactions* is dubious. Close examination of the details of PI languages shows that they generally tend to take a particular process instance to the furthest possible point of advancement before taking the advancement of any other instance into consideration. This is the avowed philosophy of GPSS, for instance (Gordon, 1975). Such exclusivity may be

As          ES
            +                        →    AS (three-phase)
            AS (time-cell)

so
            PI
            +                        →    ENG
            AS (three-phase)

Fig. 8.7 — Relationship between the ENG and other strategies. The combination of
event-driven (ES) and activity-scanning (AS) strategies gives rise to the three-phase
AS strategy, representing a compromise between temporal and configurational
preoccupations. Further combination of three-phase AS strategy with PI to form the
ENG strategy weakens the strictly sequential interpretation of the process in favour
of a more 'intelligent' and circumspect executive.

justifiable in the case where processes correspond to independent computer
programs sharing a processor, but it is certainly not true in general.

The PI strategy can be criticised for being too narrow, in its concentration on one process to the exclusion of others, thereby losing sight of the
system as a whole. It is restricted by the tramlines of its own fixed sequence,
which can lead to problems of deadlock being introduced unintentionally.
On the other hand, the AS strategy is justly criticised for being too broad on
two counts, insofar as it inefficiently tests all configurational possibilities
during the C-phase, and because any condition can serve as a C-activity test-
head. The ENG strategy seeks to combine the best of these strategies and
avoid their shortcomings.

The main benefit from the adoption of the ENG strategy is to bring out
into the open the fairness problem as it relates to discrete-event systems.
The problem, which is also mentioned in Chapter 4, is rarely brought out in
written work, but it persists in the oral traditions of simulation. An example
explained in AS terms should capture the essence of the problem. Suppose
we have two C-activities which both demand that a particular resource be in
the idle state. It is possible that, because of the C-scan order, the first
C-activity gets 'first choice' of the resource whenever it becomes available
and the second is only fulfilled when the first is satisfied. Many remedies
have been proposed, such as randomising or permuting the C-scan order, or
assigning the C-activities priorities, but none of them is widely effective and
they only push the general problem back into obscurity. Often the choice is
left to the implementation, or an *a priori* rule, such as FIFO, is blithely
applied.

Fairness is studied from a different angle in the theory of concurrent
programming (Francez, 1986). There, research is motivated by the
problems of programs being dependent on 'accidental' features of their
implementation, and the implications of these dependencies on the proof-
properties of the programs. The basic context for issues of fairness is where
there is a repetitive choice between alternatives. The situation is fair if no
alternative will be postponed forever.

Francez (1986) defines fairness as a restriction on some infinite behaviour according to the eventual occurrence of some events. He distinguishes

several kinds of fairness.

- Unconditional fairness, in which, for each behaviour, each event occurs infinitely often;
- Weak fairness, in which no event will be indefinitely postponed, as long as the process remains continuously enabled;
- Strong fairness, in which eventual occurrence is guaranteed when the process is infinitely often enabled, but not necessarily continuously.

Unconditional fairness occurs when concurrent processes are non-communicating, whereas weak fairness ensures that when a process awaits communication with another, it will eventually receive it, corresponding to a typical customer–server interaction. Strong fairness is an extension into the situation of disjunctive waiting, which may occur in allocating a general-purpose type of entity to whatever turns up next, in the area of *select* complexity, or in the field of multi-outcome waiting where an entity may accept one of a set of possible futures, depending on which happens first, as with *virtual* complexity.

What sets the concerns of concurrency theorists apart from ours is that whereas the former would like to *prove* that their systems are fair, simula-tionists are concerned with the *representation* of heuristics to arrive at decisions and to see the overall effect on the total system. Simulationists can judge fairness from the larger perspective of the 'good of the system', based on its known objective. They should receive support from the environment in which the simulation is running, which should be able to draw the simulationist's attention to the more obvious types of non-determinism.

### 8.4.3   Engagement-strategy algorithm
We present here some details of the ENG simulation executive, in which an internal net activates configurationally dependent events. An initial sugges-tion, formulated as an extension to the three-phase AS strategy to include linear flows, was first put forward by Evans (1981). Later this was further extended to provide an extra phase between C-selection and C-body-invocation (Evans, 1983).

The augmented Petri net which is used for activation control exists in the simulation program as a static data structure. Its pivotal components are the transition, place and arcs. Input to the transition are four kinds of arcs, depending on their source:

- star;
- set of facts;
- set of hard arcs;
- set of soft arcs.

If the transition is of temporal type, there is a single star arc denoting direct activation by a future event. The places input to the transition must already fully enable the transition to fire at the occurrence of the temporal event.

Control arcs may lead from facts to the transition. Each fact has an associated predicate based upon the number, type and attribute values of entities occupying the set of places which it monitors.

Hard arcs connect the transition with places which are essential for the enabling of the transition, whereas soft arcs connect places which possibly may enable it, and there may be contention between them. For instance, two arcs related by logical OR would be connected to the transition by soft arcs. Output from the transition there may be hard arcs and soft arcs connecting the transition with places, and information arcs which transmit the information that the transition has fired to places. Information arcs may be used to destroy tokens, such as is required in the case of the impatient lift-user.

To describe the activation logic, we may consider the consequences of a future-event occurrence, as summarised in Fig. 8.8. Since the occupancy requirements of input places must be automatically satisfied for a temporal transition, the transition to which the event is attached will fire without further ado. Configurational events, however, are more complicated, and a lot of checking is necessary before they can occur.

There are three types of contention: between arcs, between transitions or between tokens in a place, i.e. allocation. In many practical cases, contentions are not easily foreseen, so the simulationist may prefer to be made aware of the situation at run-time and perhaps become involved in the decision. Even if the simulationist does not choose this option, then at least *some* record should be made of the contention occurrence, before allowing the program to make an arbitrary choice. The number of contentions encountered is a measure of how loosely defined the system is.

One of the commonest situations in which contention may arise occurs in the harbour simulation of Fig. 8.5. Notice that there are two typical resources (i.e. entities composed of busy–idle cycles) in the simulation, the tug and berth. The tug is used in both *enter* and *exit* engagements and its diagram thus consists of two subcycles: *enter-idle* and *exit-idle*. An entity cycle representing a multi-purpose function will contain more than one subcycle, and has what Tocher (1964) calls a star-shaped wheel chart. It is easy to see why there is always the possibility of contention with these entities, since, for the tug becoming *idle*, there may well be ships in *wait-in* and *wait-out*, both requiring a tug, and there arises the question of what to do next.

Another problem with contention is that it can occur in unforeseeable contexts. To get the programmer to specify heuristics for all conceivable cases of contention seems a daunting task; moreover to allow the program to always stop to request guiding assistance from the experimenter would often be too disruptive. One suspects that it is probably to avoid these kinds of hindrances that the whole idea of contention, bound up with the controversy over wait-until specification, was excluded from Simula (Nygaard and Dahl, 1978). The possibility of practical user-intervention in a simulation run has had to await more interactive computer environments.

With contention dealt with, the transition fires and tokens can move from their places. The new situation causes a re-evaluation of any facts

1)   Select next future event with associated transition (F);
     WHILE there are transitions (F)
     DO
     2)   Fire transition; trigger data-probes;
          trace information arcs and invoke logic;
     3)   Remove tokens from input places;
          evaluate facts on input places;
     4)   Trace arcs to find all output places;
          mark output places (O);
     5)   Resolve *arc contention*;
          mark resolved output places (I);
          make new token engagements;
          move tokens to output places (I);
          evaluate facts on places (I);
     6)   FOR EACH place (I)
          DO
                    Trace arcs to transitions;
                    IF transition is NOT temporal
                    THEN mark transition (C)
                    FI
          OD;
     7)   FOR EACH transition (C)
          DO
                    Check firing rule for occupancy of input places;
                    Check input facts;
                    IF transition is enabled
                    THEN mark transition (E)
                    FI;
          OD;
     8)   FOR EACH transition (E)
          DO
                    Apply *allocation logic* to choose particular tokens
                          from enabling places;
                    Mark tokens (M);
                    Associate token with transition
          OD;
     9)   FOR EACH token (M)
          DO
                    IF token is associated with >1 transition THEN
                          Apply *transition contention* logic
                    FI;
                    Mark transitions (F)
          OD
     OD

Fig. 8.8 — Outline algorithm of the engagement strategy.

monitoring a place, which may lead to the enabling of further transitions. When there are no more F-transitions, the next event is taken from the FES, and the simulation time is advanced, giving rise to the opportunity for a new set of token movements. Thus the time advance proceeds, event by event, until the stopping condition, as defined by the simulationist, is reached.

From the viewpoint of control structures, the ENG strategy takes systems descriptions in the form of processes, or coroutine-templates, and converts them into procedure modules for activation by the net. The question arises: why not implement the coroutines as coroutines? In our criticism of the PI strategy, we have demonstrated the problems that too much concentration on a single entity would cause. In addition, by writing the compiled code as a collection of procedures, we also avoid the implementational problems of coroutine activation, with which there are certain difficulties (Moody and Richards, 1980; Bailes, 1985).

The ENG executive allows the possibility of coincidences of many kinds: event occurrence, contention, consequence, etc., and points them out where necessary. The algorithm is thus rather awkward, being very reluctant to allow tokens to flow far without checking for the possibility of conflict. Allocation and contention problems are thus brought to the surface for examination, rather than being subject to implementational fortune. For simplicity in the description given here, we have omitted further checks on transition firing connected with the inclusion of linear flows (Evans, 1981).

## 8.5   SIMULATION-LANGUAGE SEMANTICS

The semantics of a language should describe the *meaning* of any language construct, thereby forming a relationship between the language and its referent. It is important, then, when proposing a new language to reflect first on the domain of interest before deciding the rules of grammar. In Strachey's phrase, one should 'decide what one is going to say before worrying about how to say it' (Dana Scott in Stoy, 1977). So, we start with the semantics of SIMIAN first before looking at other issues. Marlin (1980), in his description of the coroutine language ACL, similarly leaves syntactic niceties until last.

The general problem of assigning meanings to things is quite difficult, and can be attempted in several ways. In computer science, semantics is usually discussed under one of two separate headings: the theoretical area of semantics of programming languages, and, in artificial intelligence, the network representation of relationships between data. In AI terms we need a data structure to represent the semantic descriptions of relationships to enable, say, an expert system to reason about and understand more completely statements concerning the real world. For the semantics of simulation programming languages, we find that these concerns overlap to some extent: the meaning of a language construct is already strongly bound up with a real-world representation.

There are three ways to describe the semantics of a programming language. The question: 'What does this program construct mean?' can be most simply answered by running the program on a computer and observing the effects on test data. This is the basis of the *operational* approach to defining semantics. It is unsatisfactory on two counts. Firstly, it is dependent on the particular machine used in the test. Anyone who has attempted to run a Fortran program, originally written to run on another machine, will appreciate how unsatisfactory this dependency is. Secondly, operational semantics have no external representation, which handicaps comparison and reasoning about programs. However, operational semantics are useful in discussions of models of coincident happenings in non-sequential systems (Reisig, 1984).

From the engineer's perspective of operational semantics, we jump to the opposite extreme where the semantics of a program is seen as a set of logical properties. Further properties of a program can then be deduced using axioms and rules of inference to construct a formal proof. Such *axiomatic semantics* tend to treat the program as a black box and consider meanings as mappings between input and output sequences. Are two programs semantically equivalent if for a given input they produce the same output? Clearly, there are cases where such an approach would avoid many kinds of interesting questions.

In what sense can two simulations be thought of as semantically equivalent? Behaviouristically, given the same initial configuration, they should lead to the same output, but is this enough to really capture the *meaning* of the equivalence? In fact, simulations which do not lead to the same output could still be thought of as equivalent. The changing of one random-number stream for another would not normally be thought of as a very significant change, whereas an increase in the number of resources would be highly significant.

The third approach to semantics, and probably the most actively researched, is *denotational semantics*, where we consider semantic mappings, called valuation functions, from program constructs to a domain of objects which can be regarded as being 'well known' and needing no further semantic elaboration. Every syntactically well-formed construct can then be assigned a meaning from the domain; the construct 'denotes' the domain object to which it is mapped by the valuation function. Normally the semantic domain is mathematically defined. The denotational approach has been used to define semantics of many high-level languages, such as Algol 60, Pascal and LISP, and has been used in the design of others.

For traditional algorithmic languages such as Algol, it is sufficient to base their semantic definition on mathematics, a basis which can be assumed to be 'well known'. But the types of phenomena dealt with by simulationists transcend mathematics, as we have seen in Chapter 2. Schmidt (1986) reports that other types of languages, such as the string-processing language Snobol, have received a denotational definition. But for algorithmic languages, the domain of semantic values is obviously the familiar field of mathematics: numbers, operators and sets, but for discrete-event simulation the semantic domain is not so obvious.

Complete denotational semantics have been developed for some languages which involve interacting parallel processes, usually in the form of distributed processing. The main pre-occupation of most approaches is to derive correctness proofs. Francez *et al.* (1979) offer denotational semantics for Hoare's language CSP. The semantic domain is the set of *history trees*, which are traces of inter-process communication which might have taken place, a domain first suggested by Kahn and MacQueen (1977). This work has been continued by de Bruin and Böhm (1985), who consider the mapping of input histories to output histories for dynamic networks of processes.

Park (1980), Hennessy (1984) and Francez (1986) have tried to define the semantics of fairness, as it applies in asynchronous communicating processes. The simulationist has to take a broader view of fairness, to see the system from different angles: what might be fair from one viewpoint could be unfair from another. Ultimately, there can be no 'definition' of fairness which applies to all situations across the broad application field of simulation — the point is to investigate proposed heuristics and to see their effect on some aspect of system performance.

In simulation programs, we have an abundance of procedure calls like *random*, which operate like coroutines, with their own memory — each time they are called they deliver a different result. The semantics of such constructs must take into consideration their internal state, because the value they deliver is dependent upon it. Similarly for a whole simulation, there needs to be recognition of the change in the *internal states* of a model; this same argument has been levelled against a behaviouristic approach in psychology (MacKay, 1969).

Following the second paradigm shows that the output from a simulation, though valuable, of course, is not essential to the representational aspect of languages — thus throwing out any input/output approach to semantics. If we follow the denotational approach, then it is difficult to find a domain of representation. The mathematical domains of integers, reals, etc., are all necessary as features of a simulation, as are logical predicates, but how can we relate concepts such as resource acquisition to such a domain?

Perhaps the Petri net can represent the internal states of the system, at least partially. Winksel (1980) and Nielsen *et al.* (1981) have tried to combine the concurrency of Petri nets with a denotational approach to define the semantics of communicating sequential processes. Both papers see events as a unifying concept in computation, but whether to include time in the theory, and if so, in what form, is problematic. Many papers on this topic are published in Brookes *et al.* (1985). The main motivation seems to be able to specify software for distributed processing environments.

In order to define a semantics of SIMIAN, we have chosen to define a set of semantic primitives, based on the idea of 'wait until', to describe the phenomena to be simulated. The Petri net can represent a change in the internal state of the system by the movement of a token from one place to another, and the duration of states by the persistence of tokens in places. We will consider this our 'well-known' semantic domain. The inclusion of time elapse, in the form of a star arc, is an essential augmentation of the net, enabling the representation of persistence of states. Whether the form of Petri net developed here will be useful for a more detailed semantics of simulation language is a matter for future work.

## 8.6  SIMIAN PRIMITIVES FOR ENGAGEMENTS

An uncontrolled language design, which tends to see in each new problem a new language construct, eventually leads to unnecessarily complex language features. Such a whimsical approach is probably one reason why simulation languages are often rather inelegant. On the other hand, if we adopt too rigid a standard then our descriptions may be confined by the language and we are less likely to express anything new. Generally speaking, in simulation it seems that we are still at the stage of discovering phenomena and looking for ways of describing them, so we must steer a middle course between these two extremes.

In SIMIAN we take as the basis of discrete-event simulation the wait until, in the sense that a simulation is regarded as a description of a system undergoing different kinds of wait until on behalf of many entities simultaneously. The basic description can support a higher-level and more convenient form of language, employing common assumptions and conventions. For the higher-level language, we separate the various categories of wait-until termination along the lines of our complexity characterisation, but the generality of wait until serves as a unifying concept in the semantics of the language.

A SIMIAN program is *expanded* into primitive statements which are then related to net elements. Expansion means rewriting in terms of semantic primitives. The overall plan of a SIMIAN compilation can then be described in two stages: expansion into net and transition modules, followed by the translation of these into machine code. The modularisation of the program should also facilitate incremental compilation, which is important for an effective simulation environment.

The identification of a wait-until description of discrete-event phenomena gives the possibility of mapping a simulation program into a net in which the structures of the description are abstracted. This gives a framework around which to write a compiler for the language, in which the net forms the backbone of the program. Thus the net is the central structural support for a simulation program, playing a role analogous to the parse tree of an algorithmic program.

Let us examine first the commonest high-level feature of simulation languages, the resource assumption. The language must translate the assumption into terms which would also support mutable resources, i.e. the general entity–entity interaction. For instance, the customer–server interaction in Fig. 8.9(a) and its expansion in Fig. 8.9(b).

Notice how some assumptions made in the top-level code are translated into stricter forms. Besides replacing each resource assumption by a full entity description, each engagement receives an engagement name, which is taken from the name of the distribution, if there is no clash. Otherwise the

queue:
GEN     customer;
ACQ     server;
TIME    service;
REL     server;
NEG     customer

Fig. 8.9(a) — High-level SIMIAN code for customer–server interaction.


GEN customer;
queue:          WAIT UNTIL server IN idle BY FIFO;
COOPT server FROM idle;
service:        WAIT TIME service;
FORK server INTO idle;
neg_customer:
WAIT NEG customer
***
FOR EACH server
DO CYCLE
idle:                   WAIT JOIN customer IN queue
OD.

Fig. 8.9(b) — Expansion of SIMIAN code in Fig. 8.9(a) in terms of primitives, where
*** is a process divider, and a full stop '.' denotes the end of the program.

simulationist is prompted to supply a name. The expansion shows that the main concept in the semantic model is the idea of WAIT, whether for a temporal or a configurational event. The WAIT statement indicates:

(i)  that the current entity is engaged;
(ii)  the nature of the engagement's outcome.

Each entity engagement specified in SIMIAN corresponds to a WAIT statement in the expanded version. Other primitives, such as FORK and JOIN for routing, are also necessary. For compilation, implementation from the primitive level is direct: each transition either takes apart or puts together an engagement.

Apart from the transient nature of the customer, the customer–server interaction is actually quite symmetric, although the symmetry is perhaps not readily apparent from the expansion. This is because the activity *service* in which both customer and server collaborate is written as part of the customer process. But the interaction might just as well be part of the server process. The equivalent version is given in Fig. 8.9(c). Implementations of either of these versions will have the same semantic value, and thus the same effect. Notice that the resource assumption implies a concern with the queueing discipline for the customers but not the servers, and the default discipline is FIFO.

Actually, queueing disciplines are only significant when the queueing entities are distinguishable, i.e. when they possess attributes. No entities

```
                    GEN customer;
queue:              WAIT UNTIL server IN idle BY FIFO;
                    JOIN server IN serve;
neg_customer:
                    WAIT NEG customer
***
                    FOR EACH server
                    DO CYCLE
idle:                   WAIT UNTIL customer IN queue;
                        COOPT customer FROM queue;
serve:                  WAIT TIME service;
                        FORK customer INTO neg_customer
                    OD.
```

Fig. 8.9(c) — Primitive customer–server interaction — an alternative version.

here explicitly possess attributes, but any entity created in a GENerate statement has an attribute of generation time assigned to it, in order that transit times can be calculated. Without this attribute, little of interest can be gleaned from the model. Consequently a queueing discipline is assumed. None is assumed for the servers however, unless they are given attributes by the simulationist.

Many languages draw a distinction between permanent and temporary entities. In the SIMIAN view they are distinct only insofar as their lifetimes are. So temporary entities, like the customers of this example, must be generated and terminated while permanent entities are specified in a FOR statement, which introduces their process description.

For another example, with slightly more involved engagements, we may turn to the Petri net of Fig. 8.5, the simulation of a ship entering a harbour. The top-level SIMIAN code is given in Fig. 8.10(a) and its expansion in Fig. 8.10(b).

```
                    GEN ship;
wait_in:            ACQ tug | ACQ berth | AWAIT tide IN high;
                    TIME enter;
                    REL tug;
                    TIME unload;
wait_out:           ACQ tug | AWAIT tide IN high;
                    TIME exit;
                    REL tug | REL berth;
                    NEG ship
***
                    FOR tide
                    DO CYCLE
high:                   TIME next_tide;
low:                    TIME next_tide
                    OD.
```

Fig. 8.10(a) — SIMIAN code for harbour simulation.

Once again the bulk of the expansion is made up of the details of resource processes. In this example we can see the expansion of a multi-outcome wait, specified using the | operator and the incorporation of a condition (high-tide) which must hold before the resources can be acquired.

This example also shows the emergence of the contention problem, most easily seen from the expanded version. If we consider the *tug* resource, we see that when it finishes either of its activities, *enter* or *exit*, it returns to the *idle* state whereupon it may be required by ships in one or *both* of the engagements *wait_in* or *wait_out*. As specified, the contention between these options is left unresolved and would be decided on purely implementational grounds unless some means of making the decision is specified by the simulationist at run-time. In any case, a count of contentious situations encountered would be always maintained for each run, so that the simulationist may know how 'well defined' the simulation is.

## 8.7   SIMIAN PRIMITIVES FOR ALLOCATION

Allocation, which we have so far discussed as future-event scheduling in Chapter 3, and as *select* complexity in Chapter 7, is a significant component of a discrete-event system, and therefore requires a representation in the SIMIAN language. The basic function of allocation is to *choose*, from all the different possibilities which might present themselves in a given configuration, the entity or entities to be subject to the next transition firing. Queueing discipline and attribute matching are typical allocation concerns.

Unlike the engagement sequence, allocation rules do not map naturally into a Petri net, and are therefore specified separately in program modules. Allocation modules are, however, associated with specific transitions insofar as they check the tokens on places input to the transition; allocation rules function as additional firing rules, even to the extent of inhibiting firing if the criterion is not met. For a temporal event, the allocation choice is easy enough, since each future event has its own 'priority', namely its occurrence time. All that remains is to provide an efficient algorithm to deliver the future event with minimum occurrence time, which was discussed in Chapter 3.

For configurational events, the selection of entities is based on attribute values, and can also be influenced by priority indications within the process definition itself, or even the intervention of the simulationist at run-time. Sometimes it is suggested that assigning a numerical priority to entities partaking in configurations, like assigning priorities to operators in an arithmetic expression, is sufficient to provide a selection criterion. Such a remedy is frequently pursued in the design of operating systems, where all interactions and interrupts are mediated by the respective values of priorities.

Of course in real-time applications we often do not have the luxury of being able to pursue extensive set-searching, let alone involve the user in the decision-making process. In the more relaxed area of simulation we are

```
                        GEN ship;
wait_in:        (       WAIT UNTIL tug IN idle,
                        WAIT UNTIL berth IN idle,
                        WAIT UNTIL tide IN high) BY FIFO;
                (       COOPT tug FROM idle,
                        COOPT berth FROM idle);
enter:          WAIT TIME enter;
                FORK tug INTO idle;
unload:         WAIT TIME unload;
wait_out:       (       WAIT UNTIL tug IN idle,
                        WAIT UNTIL tide IN high) BY FIFO;
                COOPT tug FROM idle;
exit:           WAIT TIME exit;
                (       FORK tug INTO idle,
                        FORK berth INTO idle);
neg_ship:       WAIT NEG ship
***

                FOR EACH berth
                DO CYCLE
idle:                   WAIT JOIN ship IN wait_in
                OD
***

                FOR EACH tug
                DO CYCLE
idle:                   (       WAIT JOIN ship IN wait_in,
                                WAIT JOIN ship IN wait_out)
                OD
***

                FOR tide
                DO CYCLE
high:                   WAIT TIME next_tide;
low:                    WAIT TIME next_tide
                OD.
```

Fig. 8.10(b) — Primitive SIMIAN code for harbour simulation.

concerned above all with the accurate portrayal of a real system, and it is not always realistic to define allocation in terms of numeric priorities; sometimes the correct decision can only be assessed from the state of the model when the event is about to occur.

The problem is basically one of scanning the total range of possibilities for entity advancement and matching attribute values against a criterion, a task essentially similar to that handled by a query-language for accessing databases, or a logic-programming language such as Prolog. Several simulation languages (e.g. SIMSCRIPT and CSL) started off by concentrating on the set-handling aspects of simulation, although this aspect of their inception is rarely emphasised nowadays.

For allocation, the first requirement of a language is to enable the definition of entity attributes. Fig. 8.11(a) gives an example of the data

```
01  ship
              02   hold VESSEL X 2
                     03   content FROM product
                     03   units Mlitres
01  tank VESSEL X 6 LIKE ship.hold
01  product DISTN
              02  SET      (a,   b,   c,   d,)
              02  DENS (0.2,  0.3,  0.4,  0.1)
01  berth X 2
01  docking DISTN
              02  SET        (1,   1.5  2.0, 2.25, 2.5, 2.75, 3.0) hrs
              02 CUMUL      (0.1, 0.15, 0.2, 0.3,  0.5, 0.7,  0.8, 1.0)
01  pipeline
              02   product
01  refinery VESSEL [ product ]
```

Fig. 8.11(a) — Data structure for tanker-discharge problem.

structure required for the tanker-discharge problem, adapted from Poole and Szymankiewicz (1977).

The data structures are defined as hierarchies suspended from named nodes. Thus we see that the *ship* has two *holds*, each of which is a standard VESSEL data type with *content* taking values from a set *product*, and *units* measured in Mlitres. Six *tanks* and a set of *refinerys* indexed by *product* are also VESSELs. The details of VESSEL will be discussed later. The *product* set is defined as {*a,b,c,d*} with an associated vector of probabilities, denoting the respective probabilities of occurrence of each product. Another probability distribution, this time cumulative, is defined in the *docking* distribution. Also defined are two *berths* and a *pipeline* with a *product* attribute.

To see how SIMIAN deals with allocation, consider the *ship* process, Fig. 8.11(b). Each *ship* is generated at a rate *lambda* and is assigned attribute values, denoting capacities and products, the latter according to a sampled value from the probability distribution specified in the *product* definition. After acquiring a *berth* resource and *docking*, the *unloading* activity starts. Each *hold* of the ship can be unloaded separately, in parallel. Firstly, a suitable storage *tank* must be found. Unfortunately, when a tank is emptied, the contents leave behind a residue which may contaminate the next batch of product. Preferably, we would choose an empty tank with the residue being the same as the hold product — otherwise, of those tanks with the same product, one with minimum contents. If there are none with the same

```
GEN   ship AT RATE lambda per hr
          hold[1] :=          ( capacity := 1600,
                      content SAMPLE product, amount := 1600 )
          hold[2] :=          ( capacity := 1600,
                      content SAMPLE product, amount := 1600 )
          await_berth:    ACQ berth:
          docking:        TIME;
          unload:         FOR EACH hold PARA
                          WHILE NOT EMPTY
                          DO
              waiting:            ACQ tank BY
                                  FOR EACH tank ARBIT IN idle
                                  WITH SAME product
                                  DO
                                      TAKE   ANY   WITH   MIN
                                          amount
                                  VOID
                                  WITH ANY product
                                      TAKE ANY IN empty
                  clean:          TIME 3 hr
                                  OD;
              connect:            TIME 45 min;
              discharge:          XFER product FROM hold TO tank
                                  AT RATE 60 per hr
                                  UNTIL tank IN FULL
                                      OR hold IN EMPTY
                                  REFX;
                                  REL tank;
              synchronise:        IF hold IN EMPTY THEN SYNC
                                      EACH hold FI
                          OD;
          undock:         TIME docking;
                          REL berth
NEG
```

Fig. 8.11(b) — The ship process, with discharge-tank selection.

product, then an empty tank can be used, but it must be cleaned first. If no tank fits these requirements, then the unloading of the hold must wait.

When *tank* and *hold* are matched, they are *connect*ed and then the discharge between *hold* and *tank* VESSELs can proceed, using a XFER (transfer) between VESSELs. A discharge activity is complete when either the *tank* is full, and another *tank* must be found, or the *hold* is empty, in which case the hold must await the completion of the discharge of all the other *holds* belonging to the same ship. To complete the model, we should attend to the transfer of product from the storage tanks to the refineries, Fig. 8.11(c).

The storage tanks and refineries are connected by a single *pipeline*, which must be cleaned between product changes. The *pipeline* process first of all acquires a *tank* which is not connected to a *hold* (i.e. in its *idle*-resource state) and has the same *product* as the *pipeline* is contaminated with. Of these, a full *tank* is chosen. If there are none, then the fullest *tank* with any *product* is taken, but the *pipeline* must be cleaned before connection.

Clearly there are many policies which could be put forward about connecting the different vessels. For instance, one might think it would be better to consider only those tanks more than half full before wasting time cleaning the pipeline. Testing out the results of different policies would be a good application of simulation.

The manipulation of entities is accomplished by a version of set processing, enabling collaboration, matching, priority-access and search to be carried out. It was the hope of von Neumann that programming languages would embrace the infinite and be able to reason with quantifiers like 'for all' and 'there exists' (Ulam, 1980). Yet the incorporation of set-theoretic concepts into programming languages seems to have lagged behind that of other features (Wells, 1980), despite the fact that it forms an important part of mathematics. One language devoted to general set operations is SETL (Schwartz *et al.*, 1986) which includes a complicated loop control structure to perform searches over sets. Its form is an extension of the conventional iterative structure of algorithmic languages. The DO–VOID–OD structure of SIMIAN combines iterative and choice control structures.

Having described how comparison between attributes of process instances is brought about, we are in a position to be more specific about the *scope* of SIMIAN identifiers, i.e. the area of program within which an identifier is recognised — its range of visibility. In algorithmic languages, scopes are naturally bounded by program modules, identifiers being brought into and taken out of scope as control passes through the modules. Alternatively, a *global* identifier may be regarded as being visible from any part of the program, or identifiers may be parcelled into blocks and their scope explicitly imported into modules, as with Fortran's COMMON blocks.

Marlin's (1980) coroutine language ACL is concerned with variables within coroutines, not their interaction in any intricate way, such as would be required to examine the attributes of a separate instance. With parameters and function values, consideration of identifier scope, or *data control*, is made explicit. In ACL and in Simula the *instance* has been the commonest basis for scope definition, and all identifier occurrences are bound to the current instance (the currently activated routine).

Marlin further makes the point that programming languages have neglected to allow identifiers to become attached to dynamic instances. Thus specific instances, be they instances of procedures invoked but not yet returned, or coroutines awaiting resumption, cannot be referred to. This lack of reference means that, for instance, deleting a node of a tree which is undergoing an overlapping scan (see Chapter 3) cannot be carried out successfully in a high-level algorithmic language, as we must check whether

FOR EACH *pipeline*
DO CYCLE
*waiting*:                    ACQ *tank* BY
                                    FOR EACH *tank* ARBIT IN *idle*
                                            WITH SAME *product*
                                    DO
                                            TAKE ANY *tank* MAX *amount*
                                    VOID
                                            WITH ANY *product*
                                            TAKE ANY *tank* MAX *amount*
                      *clean*:              TIME 45 min
          *connect*:     TIME 45 min;
          *discharge*:   XFER *product* FROM *tank* TO *refinery* [*product*]
                                VIA *pipeline*
                                AT RATE 180 per hr
                                UNTIL *tank* IN EMPTY
                         REFX;
          *disconnect*:  TIME 45 min;
                         REL *tank*
OD

Fig. 8.11(c) — Transferring the product to refineries using the pipeline.

INIT
              FOR EACH *tank*
              DO
                   *capacity* = 4000;
                   *content* SAMPLE *product*;
                   *amount* SAMPLE *uniform* (0, capacity)
              OD;
              FOR EACH *refinery*
              DO
                   *capacity* = *infinity*;
                   *amount* := 0
              OD;
              FOR EACH *pipeline*
              DO
                   *content* SAMPLE *product*;
                   INTO *waiting*
              OD
END

Fig. 8.11(d) — Initialisation segment for the tanker-discharge problem.
See section 8.9.

the node is contained in the invocation stack, i.e. whether an unreturned instance has been invoked on behalf of the node.

In simulation, it turns out that we do not really need to refer instances in isolation; therefore we do not need instance-identifiers. If every instance had an identifier, we would be forced to invent a mechanism for generating a potentially infinite set of different names for all the many entities which may be generated. Even then, the use of names would then cause the program to be too specific, and complicated.

It is clearly preferable to adopt a referencing, rather than a naming, policy to cater for the creation and destruction of temporary data areas for use by entities. The entities themselves, which flow around the net in the manner of tokens, are simply bundles of value-bearing attributes, some attached explicitly by the simulationist, some added by the run-time system. When an entity is deemed to have quit the system, it is discarded. A dynamic data-management scheme incorporating garbage collection is therefore necessary to enable re-use of the space occupied by discarded data.

In SIMIAN there are three major uses of *names*, or identifiers. Each process bears a unique name which is invoked when an entity of that name is described in a process description. We should emphasise here that the name refers to the process *template*, not to any process instances which may be derived from it. The name can be thought of as the entity *type*. Also notice that, by not requiring entities to have names, means that the rather artificial separation between type names and variable names disappears.

Within the entity structure there may be *attributes* which also bear names, to which values may be assigned. As we can see from the attribute definitions in Fig. 8.11(a), the attributes can be arranged in hierarchies. Also, there is no strict differentiation between an attribute and an entity: the attribute *hold* of *ship* takes on a life of its own as a subprocess specified within the *ship* process of Fig. 8.11(b). Besides attributes, a process possesses *engagement* names which can reveal which engagement a particular instance is currently in. These can also be arranged hierarchically, as we can see from Fig. 8.11(c) where the *clean* engagement is part of the overall *waiting* engagement.

However, we have seen in our review of simulation complexity that we frequently have to deal with collaborations between individual process instances, and often have to refer to attributes of one instance from the text of another. In making a choice from a selection set, we may demand a particular condition should hold over the whole set of candidates for the ensuing collaboration. Similarly where an entity is sharing a process with another, we need in general full access to the attribute sets of both instances.

Since entities are able to JOIN together, to jointly partake in engagements and be treated as a single entity, and be separated by FORK, it is clear that the scope of identifiers cannot be bound by program segments (i.e. processes). Within a process description we may justifiably wish to refer to identifiers belonging to other processes, as when matching the attributes of an outside entity against a criterion. The problems of scope are eased if we can assume that the *only* variables we need are entity attributes, along with those temporary hooks on which to hang the tentative findings of search loops.

If we insisted that all names be different, then we would need to invent many names, for which the counterpart in natural language would share the same name. For instance, when using programming languages with record types, it often happens that the name we give to a particular type is also the same name as we would like to give to a variable of that type. To avoid inconvenience, we can omit names, where they can be inferred from context. For instance in Fig. 8.11(b) we have the statement

> *docking*:    TIME;

where an engagement *docking* is specified to persist for a duration which is not explicit. However, *docking* is also the name of a probability distribution, so the context causes a sample to be taken from that distribution. The engagement is an abbreviation for

> *docking*:    TIME *docking*;

which seems unnecessarily repetitious. Of course, this inference can only be made if no ambiguity results: later in the same process the same distribution is sampled; in this case the engagement is explicitly named

> *undock*:    TIME *docking*;

because an engagement name must be unique within a process.

Another area where disambiguation is used in the tanker-discharge problem is with the name *product*. From Fig. 8.11(a), we can see that this name can be a distribution, a set, and an attribute of *pipeline*. The strong contextual positions of SIMIAN statements can be used to resolve the ambiguity. Furthermore, in the comparison of the attributes of different entities it would be pedantic to write

> FOR EACH *tank* ARBIT IN *idle*
>     WITH *pipeline.product* = *tank.product*

Instead, we have in Fig. 8.11(c):

> FOR EACH *tank* ARBIT IN *idle*
>     WITH SAME *product*

as the product is really the same whether it is in the tank or in the pipeline.

## 8.8   FURTHER FEATURES AND APPLICATIONS

We have seen in the, previous two sections SIMIAN facilities for engage-ments and allocations. This particular division of language function, as enshrined in the second paradigm, seems justifiable from an implementation viewpoint, but this should not prevent top-level language features from bridging the divide, if the language would benefit. The first paradigm of simulation assumes that all significant happenings are discrete; where this assumption is no longer tenable, we can cause an allocation to have a duration. In this case, we have to combine allocation with engagement inside the same linguistic construct. As an example, consider the problem of removing waiting passengers from a queue of waiting passengers and moving them inside the lift:

```
FOR EACH lift
DO CYCLE
          # at current floor . . . #
     open-door:    TIME 3;
          # wait until all pax are out, who want to be #
     unload:       WAIT WHILE
                       FOR EACH pax IN riders ARBIT
                           WITH destin = floor
                       DO
     exit:                 TAKE pax TIME 10;
                           FORK pax
                       OD;

     . . .
OD
```

Here a FOR-loop selects the lift passenger (*pax*) currently in the *riders* set, whose destination matches the current *floor*, and returns them back to their own process. It would be unrealistic to assume here that passage through the door occupies negligible time, as it is a significant contribution to the time spent at each floor. The extraction of each passenger takes ten time-units, an engagement which is regarded as subsidiary to the allocation. Meanwhile, the lift waits while the allocation is taking place, after which the lift continues its process.

Such a time-consuming allocation has a significant effect on the meaning of a FOR-loop. If we can assume that an allocation takes place in an instant of simulated time, then its parameters need only be evaluated once. If, however, allocation takes place over an extended period, then on each iteration of the loop, the parameters must be re-assigned. For an example, let us consider the loading of the lift;

```
          FOR EACH pax ARBIT IN waiting [floor]
              WHILE occupancy < capacity
              WITH SAME dir
     DO
              TAKE pax;
     enter:   COOPT pax INTO riders TIME 10
     OD
```

Once again, the movement of each selected passenger takes ten time-units, but in this situation we must allow for the possibility that more passengers will arrive in the waiting area while the lift is filling. Therefore on each iteration of the loop, the *waiting* [*floor*] set must be re-evaluated.

Versions of the FOR-loop affect both engagements and allocations. A convenient use of FOR is in *route* complexity to cause a split in an entity to pursue parallel subprocesses. If we have a set of entities, each of which requires to process on its own behalf, then a

FOR EACH *set_name* PARA DO . . . OD

construct is necessary. To specify a cyclical process, we have

FOR EACH *set_name* DO CYCLE . . . OD

Furthermore, if a collection of separately named sub-entities require their own process, a casewise processing can be specified:

FOR EACH {*sub-entity,*} CASE . . . ESAC

As an example of the last situation, consider the program segment in Fig. 8.12, describing the assembly of a chair as a part of the operations of a furniture factory, as described in Gordon (1975, p. 274). Notice that when each separate route re-unites with others, then the completed parts must wait for synchronisation. Such a high level of *route* complexity indicates the possibilities and difficulties of describing PERT networks in SIMIAN terms.

For *virtual* complexity, we once more enter several separate routes, but denote the completion of the set of parallel processes by a DROP statement. A SIMIAN version of the impatient lift-user is given in Fig. 8.13. (The Petri net is Fig. 7.5.)

The main purpose of all kinds of FOR statement is to scan all members of a given set. Whether the set is concerned with entity advancement or allocation, we will be concerned with the *order* in which the scanning takes place. By default, a set is scanned in FIFO order, as this discipline is very common, but other keywords are available, such as ARBIT, which produces a randomised order. Many simulation-language difficulties, in the expression of both engagement and allocation, are connected with manipulation of sets. SIMIAN offers several kinds of sets. We have ARRAYs for ordered sets, for which there applies *pred* and *succ* operators, and *top* and *bottom* (Fig. 8.14(a)).

Normally, discrete sets are allowed to have any number of elements, but there are some situations (e.g. lift capacity) in which a limit is necessary. This is imposed by the MAXM keyword. The SIZE keyword gives the current number of elements in the set. Attribute types can be specified as being a set OF a particular entity or a member taken FROM a set. Ultimately, the basic types are REAL, INT, BOOL or a name. For instance

STRUCT
    01  *order*
        02  *chair*
            03  *frame*
            03  *cushion* X 2
                04  *cover*
                04  *foam*
DYNA
        GEN *order* AT RATE *lambda*;
        FOR EACH *chair*
        DO
            FOR EACH *frame, cushion* PARA
                CASE
                *frame*:   ( *get_carp*:        ACQ *carpenter*;
                             *woodwork*:        TIME 55 min;
                                                REL *carpenter* ),
                *cushion*:   (FOR EACH *cushion* PARA
                             DO
                                FOR EACH *foam*, cover PARA
                                CASE
                                *foam*: (*get_fc*:   ACQ  *foam_cutter*;
                                         *cutting*:  TIME;
                                                     REL *foam cutter*),
                                *cover*: (*get_sm*:  ACQ  *sew_machine*;
                                           *sew*:    TIME;
                                                     REL  *sew_machine*)
                             *slack*: SYNC *foam* UNION *cover* AS *cushion*;
                             *assemble*:   TIME 5 min
                                ESAC;
                             OD ),
            *merge*:   SYNC *frame* UNION *cushion* X 2 AS *chair*;
            *assemble*:       TIME 20 min
                    ESAC;
            OD;
            NEG *chair*

Fig. 8.12 — Simulation of chair assembly, a component of a furniture factory.

*wait*: (                    ( *patience*:        TIME DROP;
                               *walk*:            TIME),
                           *wait_lift*:        WAIT JOIN *lift* DROP)

Fig. 8.13 — The impatient lift-user. Two separate, but linked, subprocesses are entered simultaneously, but the first to encounter the DROP statement destroys the other.

STRUCT
    01   *dir* SET (*up,down*)
    01   *floor* ARRAY (LG1, LG2, G, 1, 2, 3, 4, 5, 6, 7, 8, 9, 10)
             02   *waiting* SET OF *pax*
             02   *ext_call* [ *dir* ] BOOL
             02   *lift* SET OF *lift*
    01   *pax*
             02   *source* FROM *floor*
             02   *destin* FROM *floor*
             02   *dir*
    01   *lift* X 5
             02   *dir*
             02   *riders* SET OF *pax* MAXM 15
             02   *occupancy* = *riders*.SIZE
             02   *floor, next_floor* FROM *floor*
             02   *top, bottom* FROM *floor*
             02   *int_call* [ *floor* ] BOOL

Fig. 8.14(a) — Data structure for a lift-system simulation.

the ARRAY of floors is given names, as are the elements of the set of directions {*up, down*}. For the elements of an ARRAY, subtraction of elements gives the distance, in numbers of elements, between them.

Very often there is a great deal of symmetry in a simulation model, which the simulationist should be able to take advantage of. For instance, in the lift system, the descending lift stopping at a floor, letting passengers out and in is a mirror image of the similar situation for an ascending lift. SIMIAN provides the facility to define a *mutation range* by placing MUTA–ATUM brackets around a program segment. Within these brackets we can specify substitution alternatives inside {,} braces, as controlled by the ON condition which follows the MUTA statement. For example, Fig. 8.14(b) shows an example where a choice of operator {$<,>$} is based on the value of *dir*.

Besides sets containing discrete members, such as lifts, many predominantly discrete simulations make use of containers for substances which may be assumed to be a continuum, such as a grain silo or an oil storage tank. Although the flow to and from such containers is continuous, and should strictly speaking be simulated by a continuous model, the flow is usually constant enough to enable the discrete-event paradigms to apply. The

```
PROC call_lift (floor,pax) :
    MUTA ON dir = {up,down}
        floor.ext_call [dir] := on ;
        FOR EACH lift ARBIT
            WITH lift.floor {<,>} floor
        DO
            TAKE ANY lift WITH MIN ABS (lift.floor − floor);
                . . .
        OD
    ATUM
CORP
```

Fig. 8.14(b) — Procedure by which a passenger calls a lift.

simulation language HOCUS (Poole and Szymankiewicz, 1977) also contains facilities to handle 'continuous activities'. In SIMIAN we have the VESSEL to simulate this type of container.

Fig. 8.11 shows some examples of the use of VESSELs. Each VESSEL has the following attributes: *content*, describing the type of substance it contains; *capacity*, the maximum volume; and *amount*, the current volume of substance. Each VESSEL can exist in one of three states, referring to its current amount: *full, empty* and *partial*. In addition the vessel can be *idle*, which enables it to be ACQuired as a resource and engaged in a busy activity by an entity. Because of the special nature of continuous flow, there is a special-purpose VESSEL activity called XFER–REFX to transfer contents between VESSELs.

## 8.9  SIMIAN ENVIRONMENT

We noticed in Chapter 6 that simulations present problems to computer systems which assume the conventional division between program and data. In simulation the program (i.e. the executive) is fixed while the specific instances, i.e. descriptions of discrete-event systems for which the program is to be run, are themselves program modules of some form. It would therefore be convenient to be able to perform amendments on the system descriptions without going to the extent of re-compiling the whole simulation. This ability is especially useful when we consider the typical 'what-if' situation in which the simulationist wants to see the effect of many slightly different systems, and is probably the reason why GPSS remains an interpretative system, despite the slow running that this mode of translation entails.

To some extent, we are saved from frequent re-compilations in the SIMIAN system by the abstraction of *data-probes* from the model of the system itself. A separable experimental frame facilitates the design of the model on the one hand and design of the experimentation on the other, so that a re-run of the same model, with a different positioning of data-probes, would not incur total re-compilation.

Some types of data-probe involve identification of individual entities when they move across a transition. For instance, to collect data on the time entities spend in a queue requires the following steps.

1. At the transition where the entity enters the queue, a new attribute (a *meter* attribute) is added to the entity (transparent to the rest of the model) and assigned the value of current time.
2. At the transaction where the entity exits from the queue, the entry time recorded in the *meter* attribute is subtracted from the current time and the result stored. The *meter* attribute is then destroyed.

Entities which are being monitored for the time they spend in a state, or set of states, also require a *meter* attribute to accumulate time-in-state data. All entities created in a GEN statement have their generation time stored in a special *meter* attribute, to enable time-in-system (transit time) to be calculated.

Furthermore, each generated entity receives a unique numerical identifier, in the *id* attribute. This number is copied into any other entities which are created as offspring from this entity (along with any other attributes) to allow the matching of identifiers should the offspring ever re-emerge with their originator. The collection of data over model configurations more complex than entity-in-state can be carried out using extra facts (*meter* facts) to trigger the attribute writing and reading. They are implemented in the same way as model facts, but affect only data collection.

Overall control at run-time is in the hands of the user, interacting closely with the running program by means of instructions made through the medium of an experimental-control language, XC. Run-time control determines the length of the run, handles the insertion of data-probes, collection of statistics, causes monitoring of the various states of the model as required, and finally generates a report. This is effected by setting up *control facts* and *control attributes*. The length of a simulation run is often determined by the number of entities that have been through the system. For GPSS, this is the most common way of denoting the end of a run, through the mechanism of the *termination count*. In SIMIAN the same effect is obtained by setting up a control fact monitoring the *neg* place of an entity. Alternatively, the simulationist may wish the model run to be terminated at a particular value of *time*. This involves inserting an extra future event, a *control event*, into the FES.

In general, initialisation of a simulation is an awkward problem. A cold start, where all places are empty, is easy to specify, but is not usually very realistic. An ability to stop models and store the current model state, and perhaps to continue from that state on a later occasion, would be useful. The initialisation segment for the tanker-discharge simulation is given in Fig. 8.11(d). Similar relationships between two models must also be specified when an interrupt model is in operation. Initialisation and termination are really interrupt specifications. If a repeat run is wanted, then we should have the ability to re-start the random-numbers stream with a specific seed.

Run-time control can be thought of as a meta-simulation, with a *run-status* token flowing around a simple circular net to represent the stages of the simulation run, as in Fig. 8.15. The status token indicates the current

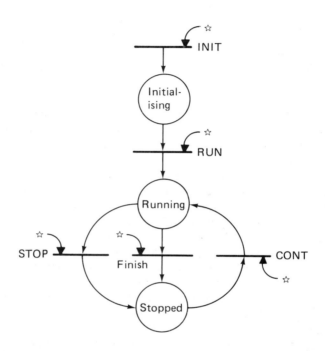

Fig. 8.15 — The net controlling the running of a simulation. The transition names are control commands issued by the simulationist. The position of a run-status token in a place gives the current run-time status of the simulation.

stage of the run. Many of the interactions between the model and the user can be mediated by *facts*. The relationship between the actual simulation and the status net is similar to that between interrupting aspects.

Another property of the control language is to define a conditional end to the simulation. For example, say we want to simulate the operations of a bank until all the customers have left the premises. Usually, banks close their doors to entering customers at the advertised closing time, but transactions may still continue on behalf of customers who are still inside the bank. Fig. 8.16 shows a net for this system, with a fact connecting the *queue* place with the run-time STOP transition.

For an effective environment, small changes, such as replacing one probability distribution by another, or adding another attribute to an entity, will incur only an *incremental* change to the program. These changes are initiated by the user editing the source code with a language-directed editor, which makes use of the symbol table constructed at the initial syntax-

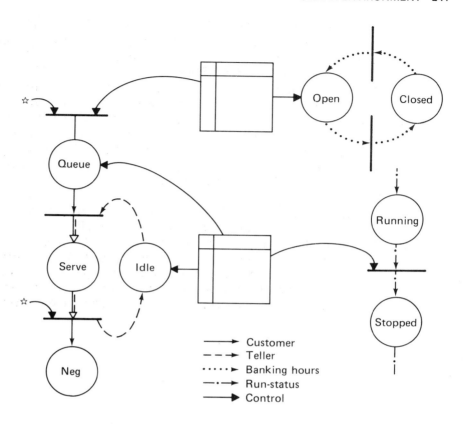

Fig. 8.16 — Simulation of bank service. The bank closes its doors at a predetermined time, but activity continues until there are no more customers in the queue and the teller is idle. This example shows the interaction between the internal simulation net and the run-time status net to define a conditional finish.

checking phase. Language-based environments are discussed by Reps (1984). Clearly, the ability to compile and re-integrate pieces of code into the main program places a constraint on the types of modules of code generated.

A SIMIAN program is compiled into C modules arranged, not on a parse tree as an algorithmic program would be, but on an augmented Petri net which is the central unifying structure for a simulation. During compilation, which comprises the normal phases of lexical and syntax analysis, a parse of the processes generates a net to act as the activating agent. Run-time debugging can be performed by marking an entity to be monitored as it passes through its processes. Running is further facilitated by a garbage collector, to scavenge through discarded objects to enable the memory they occupy to be re-used.

Having a precise statement of the model inside the computer memory should, in principle, enable a high-level dialogue between the computer and the user, in which the computer system is acting as an expert, and can reason about the system being simulated. The user should be able to supply a goal, such as 'reduce the ship turn-around time', and the expert system, by operating on the logical structure of the model, should be able to put forward recommendations for doing so.

For instance, in the tanker-discharge problem described in the previous section, an alteration in the queueing discipline to take account of the imminent arrival of ships, or of holds currently unable to find suitable tankers, would provide a basis for the improvement of the whole system. In place of the optimising simulated systems, in terms of sharpening their parameters, we envisage an *expediting* of a system to provide qualitative, strategic advice. The methodology of this aspect of the language is currently under investigation.

In addition we expect the SIMIAN system to report contention, as well as reporting on other features of the simulated system. Perhaps this could be tied in with other software for decision support. There are still many difficulties, and the language is still under development. Another simulation environment is described in Sampson and Dubreuil (1979). By abstracting the structural components of a simulation according to the second paradigm, we hope to make the running and control of simulations more effective.

# 9

# Future prospects of simulation

We consider in this chapter a few of the likely future developments in simulation. Firstly, there seems to be a promising interaction between simulation and artificial intelligence (AI), insofar as simulation will undoubtedly gain from the incorporation of expert systems and knowledge bases, and some of the SIMIAN developments in the semantic primitives and nets may be of use in describing dynamic systems for AI planning. Next we follow the common interests of AI and simulation into the area of programming paradigms and their implications in languages, environments and computer architecture designs. Finally, we close with a look at the possible impacts of simulation in future industrial applications.

## 9.1   INTELLIGENT SIMULATION

We noted in Chapter 1 the rather loose connection between simulation, computing and mental activity when we mentioned that a program could be seen as a simulated human calculation. During the development of AI the simulation of intelligent behaviour has played an influential part, although opinions differ as to the extent to which AI should be seen as the simulation of intelligence; a modern viewpoint is that AI is not really concerned with human behaviour except insofar as it may throw light on how to solve problems (Johnson-Laird and Wason, 1977). However, the simulationist can hardly fail to benefit from a more 'intelligent' programming system to aid the simulation task, and it is to this that we turn in this section.

One particular difficulty in simulating a real dynamic such as an industrial plant is that its operation frequently comprises many decision-making activities carried out by human operators, which contain estimations about imminent future happenings. Someone in control of industrial plant must frequently condone current idleness of resources in expectation of more fruitful future allocations (Amiry, 1967). To incorporate these effects, the simulationist is in the position of having to judge the operator's perception of the system, to define his short-term aim, and to decide on the appropriate choice in the given circumstances.

A more intelligent simulation language would be able to provide more assistance in this case, by being able to examine the contents of the FES. Interacting with, or controlling, a dynamic reality depends more and more on some sort of *predictive* ability about the system. When predicting, a projection into the future is made, based upon a perceived model of how the future will evolve, i.e. a mini-simulation is set up with respect to a sub-goal. Thus the goal of an adequate model includes the short-term aim of a sub-model; with increasing detail, more sub-goals may arise.

Futo and Szeredi (1982) discuss an interesting simulation of a 'goal-directed' bank robbery using interactive Prolog by means of constructing goal-seeking models, defined in terms of Horn clauses. The individual sub-goals can perhaps be seen as corresponding to the separate threads of a control heterarchy, which has been used as a basis in many AI programs (Boden, 1977). There are other approaches which may be taken in order to incorporate more intelligence into a stimulation. The framework on which a SIMIAN program is built, especially the reactivation logic incorporated into the net, can also be used as a basis for following the logical inferences manifest in a discrete-event system. An expert system may be constructed to monitor the abstract characteristics associated with discrete-event systems, such as deadlock detection, reactivation, etc. Though not necessarily 'errors' in themselves, as real systems can assume all manner of contortions, such features are frequently problematic and can be sources of error. To make simulations more realistic, further logical rules could be appended to entity descriptions to enhance the semantic content of the expert system in the particular application area. Reddy *et al.* (1986) describe the application of knowledge-based simulation in a decision-support environment, and further implications of implementing a simulation environment by using expert systems to aid the formulation of simulation problems are discussed in Balmer and Paul (1986) and Paul and Doukidis (1986).

Zisman (1978) has considered the use of Petri nets in the modelling of office procedures. The net models the control structure and the interrelation between subprocesses of the procedure. The application of nets is an interesting example of the fusion of a model of asynchronous systems with a *knowledge base*. One problem of office procedure is to know where in the procedure a certain transaction has reached, corresponding to the examination of place occupancy. Tracing the outcome of the engagements found there can help to identify a bottleneck.

How can the programming environment assist the modelling of a system with complex features? The model can be investigated for improvements in its structure, with respect to a particular goal, such as 'improve the turn-round time of aircraft'. We may envisage several kinds of situations where an expert may question the efficacy of queue discipline, for example. The expert would take the goal and apply it to the relevant entity processes and try to *expedite* each component activity. Where the activity entails a wait, then the cause of the delay could be investigated, and so on, in a top-down fashion (Evans, 1984).

To be able to handle more general intelligent activity in the object system, we must allow a wider range of possibilities to present themselves. Such programs must be heuristic in that they must respond appropriately to a general environment, or state of affairs, rather than be responsive to only a narrowly defined state. In SIMIAN, the non-linear complexity features of the language enable decisions to be reneged upon, which allows the effect of interruptive actions to be closely modelled. Without this property of being responsive to unexpected actions and being able to reduce particular conditions to general types, it is difficult to envisage any intelligent behaviour.

## 9.2   SIMULATION IN AI

Traditionally, the role of simulation in AI was to subject hypotheses to the 'rigorous logic' of the computer. Boden (1977) states the advantage of having a computer simulation of behaviour hypotheses: it is a formalisation which is activatable. Of course, simulation is still useful in this respect, but the question which interests us here is whether the mechanisms for handling discrete-event systems can have any relevance for AI and in particular, whether the type of time advance and the reactivation logic incorporated in SIMIAN may be useful in AI contexts, especially those involving *time*.

There is another perspective which interprets the future-events set as a *plan* (Sacerdoti, 1977), where each future event could be recognised as what is intended to happen at a particular time. As the planned project advances in time, future events become occurrences. But their occurrence changes the state of knowledge about the project, or changes its properties or potentialities in some way, and these changes may result in some rearrangement, e.g. cancellation, postponement, bringing forward or addition, of future-event notices.

One way in which reasoning about plans is undertaken is by means of temporal logic, which is an extension to standard logic to allow reasoning with tensed expressions (Turner, 1984). There has been some attempt to include discrete-event formalism in AI studies of planning (McDermott, 1982), who bases his temporal logic on three concepts: state, fact and event. States can be sequenced into chronicles representing the unwinding of a story, or history. Also they may represent a set of possible future courses of action, in which case the chronicles may have branched components, corresponding to decisions to be made in the future. Events may be discrete, which means they never overlap, or they may possess duration (and then correspond to what we call 'activities'). We have seen how a short-term happening, such as walking through a door, may be described either by a discrete event or by an activity with duration, which can be thought of 'expanding' a transition to include a durational place in the net.

McDermott (1983) introduces a discrete-event formalism for reasoning about plans, where events are regarded as discrete only if they can be isolated. A basic assumption in planning is that the truth value of a statement is time-dependent. One problem is that if we stay with two-valued logic, i.e. admit that facts are strictly only either true or false, then we have trouble with plans being undertaken where their goal is 'becoming' true, but is not true yet. The properties of McDermott's temporal plans can all be described by augmented Petri nets and the complexities discussed in Chapter 7.

In early work, planners conveniently assumed that actions take place instantaneously. Temporal relations are therefore restricted to 'before', 'after' and 'in parallel', limiting the expression of more complex goals like aiming to finish two separate activities at the same time. Allen and Kautz (1983) present a view of a plan as a collection of assertions viewed as an abstract simulation of some future world, including actions, events and states, bound up in a causal nexus. Allen and Koomen (1983) and Allen (1983) present a general temporal scheme based on time intervals, in which relations can be changed as time proceeds. Tsang (1986a) situates Allen's logic within other temporal logics and seeks to modify it for planning under uncertainty of future happenings, and goes on further (Tsang, 1986b) to develop a planner based on an interval-time logic.

Turner (1986) notes a similarity between the techniques of planning and simulation when deducing the state of a particular point in time. In planning, each scheduled event-notice is recognised as a contingency from the outset, rather than, in the case of simulation, as a certainty unless cancelled. This small change of emphasis, together with some ability to reason with the semantics of inter-event relationships, means that in planning we require a closer interaction with the FES itself.

In simulation, the future is always uncertain, until it becomes the present, and then passes irreversibly into certainty. A future-event notice becomes an actual event occurrence only when it is the uniquely most imminent event-notice in the FES. Until that situation holds, any future-event notice is prone to cancellation, especially when we are dealing with engagement which may be subject to interrupt.

The SIMIAN primitives when expressed in terms of the basic wait-until semantics express a system in terms of a state of readiness *for* a future happening. Net abstraction is at the level of reactivation *possibility*, and all future events are cancellable and all engagements are interruptible, so the system is never inevitably committed to any future happening. Furthermore, the forms of non-linear complexity enable us to describe a variety of possible future developments, in the manner of McDermott's branching chronicles. In Drummond (1986), planning systems are proposed nearer to the approach taken in SIMIAN by making use of a Petri net, but without making use of net predicates (facts) as a source of reactivation.

## 9.3   PROGRAMMING PARADIGMS

We have already remarked in Chapter 3 the difficulties of making use of Algol 68 as a simulation language. Generally speaking, any algorithmic language will present the same problems: the need to specify procedures at run-time, to use abstract data types, to define operations without specifying implementational details, etc. Many of these requirements are now being considered in the development of *object-oriented* programming languages.

Simula (Birtwistle *et al.*, 1973), based on Algol 60, was the first language to embody the object concept, as we have noted in Chapter 6. Today it would be called Object Algol. Its notion of *type* includes classes whose instances may be assigned as values of class-valued variables and may persist between the execution of the procedures they contain. Procedures and data declarations of a class constitute its interface and are accessible to users. Simula's *classes* are user-defined types organised in a simple inclusion (or inheritance) hierarchy in which every class has a unique immediate super-class. Simula's objects and procedures are *polymorphic* in the sense that an object of a subclass can appear wherever an object of the superclass is required.

Subsequent object-oriented languages like Smalltalk-80 combine the class concept derived from Simula with a stronger notion of information hiding, especially useful for programming in the large. As well as providing new concepts for programming simulations, the object-oriented paradigm has affected the way in which the programming environment relates to programs. Riddle and Sayler (1979) describe an object-oriented environment for modelling and simulation in the design of complex software systems. A good introduction to the software-engineering aspects of object-oriented programming is given in LaLonde *et al.* (1985).

Cardelli and Wegner (1985) have since defined object-oriented concepts in a more general manner, in order to compare the properties of later object-oriented languages. A language is object-oriented if it supports objects that are data abstractions with an interface of named operations and a hidden local state. Objects may have an associated object type, which may inherit attributes from supertypes. In general, object-oriented languages may be characterised in many ways: the principal property being that the object definition is distributed among facets of its behaviour, enabling an interface between program and object. Data abstraction provides an alternative to conventional procedure-oriented programming. For general-purpose programming, the main aim is to encourage modularity, and by this means ease development problems of large systems.

Ada is a language with facilities for data abstraction, and it has been used to write simple PI-strategy simulations (Tellaeche Bosch and Downes, 1983), implementing entity processes as Ada tasks. However, the separation of the general from the specific parts of the simulation is not so straightforward. While Ada tasks and generics hold great promise in the representation of parallel processes, there are some problems in the language's fulfilling these requirements. For instance, the attributes of a task are inaccessible to other tasks. Embodying the general simulation facilities in a package also runs into difficulties (Unger *et al.*, 1984). Despite its extensive properties, ·Ada is not a true object-oriented language (Kreutzer, 1986).

Nygaard (1986) gives a generalised object-oriented perspective over the whole science of informatics. An information process is a system developing through its state transformations, performed as actions by objects. Three important programming perspectives are functional-orientation, object-orientation and constraint-orientation. Each of these views of the computing process should be supported, in the opinion of the author, rather than hoping that one particular perspective will dominate the others.

Other programming paradigms have been suggested for writing simulations. We saw in Chapter 8 the particular problem involved in specifying a choice during an allocation, and that in general programming terms this turns out to be a set-handling problem. In fact, both CSL and SIMSCRIPT started out from a set-handling perspective. Franta and Maly (1975) have attempted to interpret simulation phenomena in terms of the set-handling language SETL.

The parallel between the development in time of a simulation and the progression through a sequence of logical steps in a problem-solving program has been referred to (Vaucher, 1985). However, the means of progression in each case is quite different: in the simulation case the system is driven by event occurrence, whereas problem solution attempts to strive towards a goal-constituted final state. Backtracking and heuristics, common features of logical programming, have no direct counterpart in conventional approaches to simulation. Various concurrent versions of Prolog have been used to describe discrete systems, but it is felt that more study is required to assess the usefulness of this connection (e.g. Cleary *et al.*, 1985).

An early application of simulation in terms of logic is the development of T-Prolog (Futo and Szeredi, 1982). The language is an extension of Prolog to offer the facilities of familiar simulation languages in a logic-programming context. As we have seen, using the facilities of backtracking, a goal-directed behaviour can be simulated by having the model automatically altered until it exhibits some desired property.

The logic-programming language Parlog (Broda and Gregory, 1984) can support a PI-strategy simulation by means of a 'communicating processes' approach in which synchronisation, updating and testing of state are achieved by message-passing between processes; processes are modelled by Parlog relations and communication channels by Parlog streams. The mode of representing a real-world model by communicating processes lends itself to a diagrammatic means of expression, where each process is a node, and the channels are directed arcs, and the diagram can be interpreted as a Parlog program in an alternative syntax. The implementational details (FES, *hold* procedure, etc.) can also be represented in the same form.

## 9.4 DISTRIBUTED SIMULATION

A discrete-event simulation program written according to the PI strategy superficially resembles a distributed program with separate processes working away almost independently, occasionally interacting with one another. Many introductions to simulation stress that simulation is about representing a parallel reality on sequential hardware, and one is left wondering whether, with the advent of parallel hardware, all this concern with time advance, strategy, even complexity might suddenly be resolved, leaving the concerns of this book an interesting anachronism.

We have remarked before that the main problem for simulation languages is the sequentialisation of a system of parallel entity flows. But if we consider the target machine of the simulation code as a machine with parallel components itself, how can we best match up the parallelism of the model with that of the machine? This is the central question posed in implementing simulations on distributed hardware.

A complex simulation system can be subdivided into parallel-running components in many ways. The separate tasks — random-number generation, statistical analysis, etc. — may be assigned to separate processors. A model-based mapping associates, on a one-to-one basis, the components of the model with corresponding processors. The main problem is to ensure synchronisation between the coordinating processors. Information about the occurrence of an event in one processor could be broadcast by a message passed between processors. Zeigler (1985) describes the application of his DEVS formalism for a discrete-event system to the problem of mounting a simulation program on distributed hardware so as to use the natural parallelism inherent in the model, by mapping the DEVS model onto a physical hierarchical network of multi-processors.

Jefferson and Sowizral (1985) propose a *time-warp* mapping in which simulation is treated as a distributed programming problem, i.e. as a set of processes advancing in parallel. In the time-warp approach each process advances independently, only stopping to synchronise when necessary, and employing roll-back if availability assumptions made previously prove mistaken. We may typify this approach as 'lazy synchronisation with roll-back'. However, a closer examination of the ideas behind this scheme show that the relationship between distributed programming and simulation is only skin-deep.

In the first place, different interpretations are put on the word 'process'. For distributed programming, the processes seem to co-exist for the duration of the run; they are created at the start of the program and all persist to the end. This assumption does not fit in well with simulation, where we prefer to think of a process as a definitional template from which process instances are generated, either at simulation origination, or from time to time during the simulation run. Process instances are continually being created and destroyed, according to the requirements of the simulation, and furthermore they can enter into relations of dependency with one another, such as those described by non-linear complexity: submission, branching (both real and virtual), merging and interruption. Some kind of global mechanism seems necessary to control and oversee the interactions of simulation processes.

Secondly, the first paradigm of simulation does not apply to distributed programming. In simulation, when an entity changes state, it does so at an instant of simulated time. Between such instants, corresponding to the time spent in engagements, the entity remains constant and therefore does not require any computation. This basic fact allows simulation to hop from one event to the next in the event-to-event time advance. Distributed processes, on the other hand, are continually processing their own computations, stopping only occasionally to attend to synchronisation with their fellows. What they do between stops is definitely not ignorable.

Thirdly, to extrapolate likely applications and environments that will be taken up by simulation in the future, more intense interaction with the simulationist is likely to become increasingly significant. Increased interaction is concomitant with closer integration of simulation with decision-support facilities. User interaction would be very confusing if there were no synchronised global time value. Decisions made once might be reneged upon, leaving the user with an utterly chaotic impression.

Distributed simulation can be achieved by bolting together what are essentially von Neumann machines in parallel; another approach is to make use of the parallelism inherent in the problem, as represented by the second paradigm. The process describes a sequence of engagements; owing to the second paradigm, the processes are nothing but the empty hulls of engagement assignments and entity routing; the allocations may be carried out by a separate control flow; and we have already seen how the data-probes can be inserted separately to perform data collection, which would be extraordinarily complicated if the processes were to run according to the time-warp.

Treleaven *et al.* (1982) discuss three essentially different kinds of computation: data-driven, demand-driven and control flow. Applying the second-paradigm breakdown of control flow for a simulation program, we find that data-driven computation is what we require to implement data-probes, which are separate threads of logic invoked by control passing through a certain point (i.e. the transition), while the main flow of control is the token flow through the net. The future of computer architecture, if it is to remain general-purpose, must be through some recognition of different kinds of architectural designs, interacting in parallel.

The motivation for the engagement strategy is based on the difficulties of implementing PI simulations in a fair way, and is therefore essentially a criticism of the single thread of instructions inherent in von Neumann architecture. Simulation is about parallelism in its *most general* sense: it has the complexity of the whole field of designed systems to handle, and will therefore continue to play an innovative role in testing out new architectural proposals.

## 9.5  FUTURE APPLICATIONS

Simulation started as a very direct means of solving industrial production problems by experiment ('what-if') without actually committing any real production. Since those days we have seen the development of simulation software, and through languages, methodologies and environments. The applications have led to new control structures and have brought about an abstract structure general enough to describe any discrete-event phenomena. The next stage seems to be the re-application of techniques to new production problems caused by the advent of new manufacturing systems.

Computer-integrated manufacturing (CIM) has been made possible by the flexibility which can be built into machines containing microprocessor chips, which means that, by using a flexible manufacturing system (FMS), a manager may swiftly re-tool and re-configure his manufacturing plant to make whatever he finds appropriate from among his product range. Of course, there will be many kinds of indirect constraints on his choice: personnel, finance, stocks, work-in-progress, the market, etc., but a well-integrated system can consider all the relevant aspects together in a simulation, to enable a wise choice to be made.

Coll *et al.* (1985) consider the application of simulation to (industrial) production systems, especially in regard to flexible manufacturing systems and CIM. Simulation has two main roles: to evaluate a production system design and as an online control mechanism, or decision-support system. The latter role would allow the possibility of total CIM, involving simulation in an essential way, through an integration of computer-based automation and decision-support systems, leading to a total control of the manufacturing system: product design, manufacture and distribution, including production, inventory and financial management. The prospect creates the possibility of the unmanned factory. Recently, several conferences have been devoted to simulation in manufacturing (Hurrion, 1986; Lenz, 1986).

Although simulation is already successful in the field of production and service planning, there are some drawbacks in its use. These are mainly confined to what we might call the *initialisation problem*. Given a particular arrangement of investment, work-in-progress, a manager might want to quickly ask a 'what-if' question of the simulation as to whether he should go ahead on a new venture or not. Owing to the traditional subdivisions between disciplines, data on the current state of affairs would normally be under the aegis of a different package, probably a database of some kind. In other words, even though all management information might be computerised, it may require several 'journeys' to the computer in order to get the initial data together. True decision support must be able to override such divisions in our thought patterns, which only arise from the historical, accidental approach in which the field has evolved.

Simulation as a decision tool will undoubtably become more valuable in production planning, especially as production becomes more automated and flexible. Seen as part of a decision-support system, simulation will transcend the traditional OR/MS or MIS role, as a more incisive and tactical decision-maker. And insight, at first jeopardised in the turning away from the strictures of the statistical model (Ignall and Kolesar, 1979), will be regained in a more qualitative and specific form.

# Discussion

Some ideas are offered for use in class discussion, exercises or projects. They are numbered by chapter.

1 Consider the fivefold modelling categorisation of Haggett and Chorley (1967) and use it to classify the following:

- a banknote;
- a dramatic production;
- an identity card;
- a fashion photograph;
- the simulation paradigms.

Many objects will show characters of more than one category.

2 A chemical reaction can be thought of as a meeting of discrete entities — try to formulate a reaction as a discrete-event system.

3 Invent new methods of indexing linear lists, and partial-balance mechanisms for trees and compare them with other methods of implementing FES operations.

4(i) Why would it be unwise to allow *read* statements inside event routines?

4(ii) Compare coroutine operation with:

(a) input-output programming;
(b) multi-pass compiling;
(c) communication with operating system during a batch job.
(see Knuth (1973a), section 1.4.2)

4(iii) In the following models construct:

(a) entities, attributes, resources;
(b) list of events, activities, processes;
(c) define event-modules, B- and C-activities;
(d) data-probes for collecting pertinent data.

- road traffic junction;
- doctor's waiting room;
- boarding an international flight;
- university admissions system.

4(iv) Investigate the road traffic flow at a neighbourhood traffic junction. Describe the types of phenomena observed and write a simulation to reproduce these phenomena. How can their prevalence be measured?

5 Take a chaotic generator and test its output sequence for satisfying the requirements of randomness.

6 Take any simulation language and consider its design. What strategy does it take? Does it supply enough facilities? How fair is it in allocation?

7 Take a large simulation example from a compendium. To what extent does the sevenfold classification of complexity cover the difficulties involved?

8 Design the following SIMIAN language facilities:

- a search control structure;
- facilities for activity network simulation;
- facilities for direct simulation from Petri nets.

9 Consider the likely social and political effects of automated production plants.

# References

Adam, N. R. and Dogramaci, A. (Eds) (1979) *Current Issues in Computer Simulation*, Academic Press, New York.

Adel'son–Vel'skii, G. M. and Landis, E. M. (1962) 'An Algorithm for the Organisation of Information', *Soviet Mathematics* **3** 1259–1263. A translation from *Doklady Akademia Nauk SSSR* **146** (1962) 263–266.

Aho, A. V., Hopcroft, J. E. and Ullman, J. D. (1974) *The Design and Analysis of Computer Algorithms*, Addison-Wesley, Reading, Mass.

Aho, A. V., Hopcroft, J. E. and Ullman, J. D. (1983) *Data Structures and Algorithms*, Addison–Wesley, Reading, Mass.

Aho, A. V. and Ullman, J. D. (1977) *Principles of Compiler Design*, Addison-Wesley, Reading, Mass.

Andrews, G. R. and Schneider, F. B. (1983) 'Concepts and Notations for Concurrent Programming', *ACM Computing Surveys* **15** 3–43.

Allen, J. F. (1983) 'Maintaining Knowledge about Temporal Intervals' *Comm. ACM* **26** 11 832–843.

Allen, J. F. and Kautz, H. A. (1983) 'A Model of Naive Temporal Reasoning', in Hobbs and Moore (1983) 251–268.

Allen, J. F. and Koomen, J. A. (1983) 'Planning using a Temporal World Model', *Proc. IJCAI-83*, Karlsruhe, 741–747.

Allison, L. (1983) 'Stable Marriages by Coroutines', *Information Processing Letters* **16** 61–65.

Amiry, A. P. (1967) 'The Simulation of Information Flow in a Steel-making Plant', in Hollingdale, S. H. (Ed) (1967) 157–165.

Aris, R. (1978) *Mathematical Modelling Techniques*, Pitman, San Francisco.

Åström, K. J. and Wittenmark, B. (1984) *Computer Controlled Systems*, Prentice-Hall, Englewood Cliffs, NJ.

Babich, A. F., Grason, J. and Parnas, D. L. (1975) 'Significant Event Simulation', *Comm. ACM* **18** 6 323–329.

Bailes, P. (1985) 'A Low-cost Implementation of Coroutines for C', *Software — Practice and Experience* **15** 4 379–395.

Balmer, D. W. and Paul, R. J. (1986) 'CASM — The Right Environment for Simulation', *Journal of the Operational Research Society* **37** 5 443–452.

Barnes, J. G. P. (1976) *RTL/2 Design and Philosophy*, Heyden, London.

Barron, D. W. (1977) *An Introduction to the Study of Programming Languages*, Cambridge University Press, Cambridge.

Bartlett, M. S. (1975) *Probability, Statistics and Time*, Chapman and Hall, London.

Bent, M. E. (1976) *Pilot GSP IV Manual*, British Steel Corporation, Birmingham.

Berry, D. M. (1981) 'Remarks on R.D. Tennent's Language Design Methods Based on Semantic Principles: Algol 68, a Language Designed Using Semantic Principles', *Acta Informatica* 15 83–98.

Birtwistle, G. M. (1979) *Discrete Event Modelling on Simula*, Macmillan, Basingstoke.

Birtwistle, G. M., Dahl, O.-J., Myhrhaug, B. and Nygaard, K. (1973) *'SIMULA BEGIN'*, Studentlitteratur, Lund.

Birtwistle, G., Lomow, G., Unger, B. and Luker, P. (1985) 'Process Style Packages for Discrete Event Modelling: Experience from the Transaction, Activity and Event Approaches', *Transactions of the Society for Computer Simulation* 2 1 27–56.

Blackstone, J. H., Jr., Hogg, G. L. and Phillips, D. T. (1981) 'A Two-list Synchronization Procedure for Discrete Event Simulation', *Comm. ACM* 24 12 825–829.

Blunden, G. P. (1968) 'Implicit Interaction in Process Models', in Buxton (1968), 283–291.

Boden, M. A. (1977) *Artificial Intelligence and Natural Man*, Harvester Press, Brighton.

Bohm, D. (1980) *Wholeness and the Implicate Order*, Routledge and Kegan Paul, London.

Bornat, R. (1979) *Understanding and Writing Compilers*, Macmillan, London, Chapter 14.

Box, G. E. P. and Muller, M. E. (1958) 'A Note on the Generation of Random Normal Variates', *Ann. Math. Stat.* 29 610–611.

Boyer, C. B. (1968) *A History of Mathematics*, Wiley, New York.

Bratley, P., Fox, B. L. and Schrage, L. E. (1983) *A Guide to Simulation*, Springer-Verlag, New York.

Brecht, Berthold (1957) *Versuche*, Band II, S.104, Frankfurt.

Brennan, R. D. (1968) 'Continuous System Modelling Programs: State-of-the-Art and Prospectus for Development', in Buxton, J. N. (Ed.) (1968), 371–396.

Brinch Hansen, P. (1977) *The Architecture of Concurrent Programs*, Prentice-Hall, Englewood Cliffs, N.J.

Broda, K. and Gregory, S. (1984) *Parlog for Discrete Event Simulation*, Research Report DOC 84/5, Department of Computing, Imperial College, London.

Brookes, S. D., Roscoe, A. W. and Winskel, G. (Eds) (1985) *Seminar on Concurrency*, Lecture Notes in Computer Science, 197, Springer-Verlag, Berlin.

Brown, M. R. (1978) 'Implementation and Analysis of Binomial Queue Algorithms', *SIAM Journal of Computing* 7 3 298–319.

de Bruin, A. and Böhm, W. (1985) 'The Denotational Semantics of Dynamic Networks of Processes', *ACM Trans. on Programming Lan-*

*guages and Systems* **7** 4 656–679.

Buxton, J. N. (1966) 'Writing Simulations in CSL', *Computer Journal* **9** 137–143.

Buxton, J. N. (Ed.) (1968) *Simulation Programming Languages*, Proceedings of the IFIP Working Conference, Oslo, May 1967, North-Holland, Amsterdam.

Buxton, J. N. and Laski, J. G. (1962) 'Control and Simulation Language', *Computer Journal* **5** 3 194–199.

Cadzow, J. A. (1973) *Discrete-time Systems*, Prentice-Hall, Englewood Cliffs, NJ.

le Calvez, F., Mendelbaum, H. G. and Madaule, F. M. (1977) 'Compiling GAELIC: a Global R-T Language', in Smedema, C. H. (Ed.) *IFAC/ IFIP Workshop on R-T Programming*, North Holland, Eindhoven.

Campbell, J. A. (1984) *Three Uncertainties of Artificial Intelligence*, Research Report W120, University of Exeter.

Cardelli, L. and Wegner, P. (1985) 'On Understanding Types, Data Abstraction, and Polymorphism', *ACM Computing Surveys* **17** 4 471–522.

Carrie, A. S., Adhami, E., Stephens, A. and Murdoch, I. C. (1985) 'Introducing a Flexible Manufacturing System', in Rauof, A. and Ahmed, S. I. (Eds) *Flexible Manufacturing: Recent Developments in FMS, Robotics, CAD/CAM, CIM*, Elsevier, Amsterdam, 1–13.

de Carvalho, R. S. and Crookes, J. G. (1976) 'Cellular Simulation', *Operational Research Quarterly*, **27** 1 31–40.

Cellier, F. E. (1979) 'Combined Continuous/Discrete System Simulation Languages — Usefulness, Experiences and Future Development' in Zeigler, B. P. *et al.* (1979) 201–220.

Chaitin, G. (1975) 'Randomness and Mathematical Proof', *Scientific American* **232** 5 47–52.

Chatfield, C. (1980) *The Analysis of Time Series*, Second Edition, Chapman and Hall, London.

Checkland, P. (1983) 'O. R. and the Systems Movement: Mappings and Conflict', *Journal of the Operational Research Society* **34** 8 661–675.

Churchman, C. W., Ackoff, R. L. and Arnoff, E. L. (1957) *Introduction to Operations Research*, Part III, Wiley, New York.

Clark, R. N. (1985) 'A Pseudorandom Number Generator', *Simulation* **45** 5 252–255.

Cleary, J., Goh, K.-S. and Unger, B. (1985): 'Discrete Event Simulation in Prolog', Proc. SCS Conference on AI, Graphics and Simulation, San Diego, January 1985, 8–13.

Clementson, A. T. (1966) 'Extended Control and Simulation Language', *Computer Journal* **9** 3 215–220.

Clementson, A. T. (1973) *Extended Control and Simulation Language, Users Manual*, Institute for Engineering Production, University of Birmingham.

Clementson, A. T. (1977) *Extended Control and Simulation Language — Computer Aided Programming System*, Lucas Institute for Engineering

Production, University of Birmingham.

Coll, A., Brennan, L. and Browne, J. (1985) 'Digital Simulation Modelling of Production Systems', in Falster and Mazumder (Eds) (1985) 175–195.

Comfort, J. C. (1979) 'A Taxonomy and Analysis of Event Set Management Algorithms for Discrete Event Simulation', *Proc. Twelfth Annual Simulation Symposium*, Tampa, Florida, 115–146.

Conolly, B. (1981) *Techniques in Operational Research*, 2 'Models, Search and Randomization', Ellis Horwood, Chichester.

Conway, M. E. (1963) 'Design of a Separable Transition-diagram Compiler', *Comm. ACM* 6 396–408.

Conway, R. W., Johnson, B. M. and Maxwell, W. L. (1959) 'Some Problems of Digital Systems Simulation', *Management Science*, 6 92–110.

Cox, D. R. and Miller, H. D. (1965) *The Theory of Stochastic Processes*, Methuen, London.

Crane, C. A. (1980) *Linear Lists and Priority Queues as Balanced Binary Trees*, Garland, New York; Ph.D. dissertation, Stanford University, 1972.

Crookes, J. G., Balmer, D. W., Chew, S. T. and Paul, R. J. (1986) 'A Three-phase Simulation System Written in Pascal', *J. Opl Res. Soc.* 37 6 603-618.

Crosson, F. J. and Sayre, K. M. (1963) 'Modeling: Simulation and Replication', in Sayre, K. M. and Crosson, F. J. (1963) *Modeling of Mind, Computers and Intelligence*, Simon and Schuster, New York.

Dahl, O.-J. (1968) 'Discrete Event Simulation Languages', in Genuys, F. (Ed.) *Programming Languages*, Academic Press, London, 349–395.

Dahl, O.-J. and Nygaard, K. (1966) 'SIMULA — an Algol-based Simulation Language', *Comm. ACM* 9 9 671–678.

Davey, D. and Vaucher, J. (1980) 'Self-optimising Partition Sequencing Sets for Discrete Event Simulation', *INFOR* 18 21–41.

Davies, N. R. (1979) 'Interactive Simulation Program Generation', in Zeigler, B. P., *et al.* (1979) 179–200.

Delfosse, C. M. (1976) *Continuous and Combined Simulation in SIMSCRIPT II.5*, CACI Inc., Arlington, Va.

Deo, N. (1979) *System Simulation with Digital Computer*, Prentice-Hall of India, New Delhi.

Dijkstra, E. W. (1968a) 'Co-operating Sequential Processes' in Genuys, F. (Ed.) *Programming Languages*, Academic Press, London, 43–112.

Dijkstra, E. W. (1968b) 'Go To Statement Considered Harmful', *Comm. ACM* 11 3 147–8.

Dijkstra, E. W. (1968c) 'The Structure of the "THE" Multiprogramming System', *Comm. ACM* 11 8 341–349.

Dijkstra, E. W. (1975) 'Guarded Commands, Nondeterminacy and Formal Derivation of Programs', *Comm. ACM* 18 8 453–457.

Dijkstra, E. W. (1976) *A Discipline of Programming*, Prentice-Hall, Englewood Cliffs, NJ.

Dorn, W. S. and McCracken, D. D. (1972) *Numerical Methods with*

*Fortran IV Case Studies*, Wiley, New York.

Drummond, M. E. (1986) *A Representation of Action and Belief for Automatic Planning Systems*, Report AIAI-TR-16, Artificial Intelligence Applications Institute, University of Edinburgh.

Edwards, R. (1977) 'Is Pascal a Logical Subset of Algol 68 or Not?', Proc. Strathclyde Algol 68 Conference, *ACM Sigplan Notices* **12** 6 (June) 184–191.

Ellison, D. (1979) 'A Combined Executive for Activity Based Language (CABL)', Winter Computer Simulation Conference, Toronto, 77–79.

Elmaghraby, S. E. (1977) *Activity Networks*, Wiley, New York.

Engelbrecht-Wiggans, R. and Maxwell, W. L. (1978) 'Analysis of the Time Indexed List Procedure for Synchronization of Discrete Event Simulations', *Management Science* **24** 13 1417–1427.

Enslow, P. H. (Ed.) (1974) *Multiprocessors and Parallel Processing*, Comtre Corporation.

Evans, J. B. (1981) 'Discrete Event Simulation Package for Modelling Entities with Many Aspects — a Reappraisal of the Activity Approach', in *Proc. Summer Computer Simulation Conference*, Washington, DC, 58–61.

Evans, J. B. (1983) *Investigations into the Scheduling of Events and Modelling of Interrupts in Discrete Event Simulation*, Ph.D. thesis, Department of Operational Research, University of Lancaster.

Evans, J. B. (1984) *Simulation and Intelligence*, Technical Report TR-A5-84, Centre of Computer Studies and Applications, University of Hong Kong.

Evans, J. B. (1986a) 'A Characterisation of Discrete-event Simulation Complexity', *JSST Conference on Recent Advances in Simulation of Complex Systems*, Tokyo, 1986, 24–28.

Evans, J. B. (1986b) 'Experiments with Trees for the Storage and Retrieval of Future Events', *Information Processing Letters* **22** 237–242.

Fahrland, D. A. (1970) 'Combined Discrete Event/Continuous Systems Simulation', *Simulation* **14** 2 61–72.

Falster, P. and Mazumder, R. B. (Eds) (1985) *Modelling Production Management Systems*, North-Holland, Amsterdam.

Favrel, J. and Lee, K. H. (1985) 'Modelling, Analyzing, Scheduling and Control of Flexible Manufacturing Systems by Petri Nets', in Falster and Mazumder (1985) 223–243.

Feyerabend, P. (1978) *Against Method*, Verso, London.

Finlay, M. I. (1970) *Early Greece: the Bronze and Archaic Ages*, Chatto and Windus, London.

Fishman, G. S. (1978) *Principles of Discrete Event Simulation*, Wiley, New York.

von Foerster, H. (1980) 'Epistemology of Communication', in Woodward, K. (Ed.) *The Myths of Information: Technology and Postindustrial Culture*, Coda, Madison, WI.

Foster, C. C. (1973) 'A Generalisation of AVL Trees', *Comm. ACM* **16** 8 513–517.

Francez, N. (1986) *Fairness*, Springer-Verlag, New York.

Francez, N., Hoare, C. A. R., Lehman, D. J. and de Roever, W. P. (1979) 'Semantics of Nondeterminism, Concurrency and Communication', *J. Comp. and Sys. Sci.*, **19** 3 290–308.

Françon, J., Viennot, G. and Vuillemin, J. (1978) 'Description and Analysis of an Efficient Priority Queue Representation', *Proc. 19th Annual Symposium on the Foundations of Computer Science*, Piscataway, NJ. 1–7.

Franta, W. R. (1977) *The Process View of Simulation*, North Holland, New York.

Franta, W. R. and Maly, K. (1975) 'Simulation Structures and SETL', *Information Processing* **74**, North Holland, Amsterdam.

Franta, W. R. and Maly, K. (1977) 'An Efficient Data Structure for the Simulation Event Set', *Comm. ACM* **20** 8 596–602.

Franta, W. R. and Maly, K. (1978) 'A Comparison of Heaps and the TL Structure for the Simulation Event Set', *Comm. ACM* **21** 10 873–875.

Fredman, M. L., Sedgewick, R., Sleator, D. D. and Tarjan, R. E. (1986) 'The Pairing Heap: a New Form of Self-adjusting Heap', *Algorithmica* **1** 111–129.

Futo, I. and Szeredi, J. (1982) 'A Discrete Simulation System based on Artificial Intelligence Methods', in Javor, A. (Ed.) *Discrete Simulation and Related Fields*, North-Holland, Amsterdam.

Gardner, M. (1970) 'The Fantastic Combinations of John Conway's New Solitaire Game "Life"', *Scientific American* **223** 4 120–123.

Gardner, M. (1971) 'On Cellular Automata, Self-reproduction, the Garden of Eden and the Game "Life"', *Scientific American* **224** 2 112–117.

Gardner, M. (1982) *Logic Machines and Diagrams*, Second Edition, University of Chicago Press, Chicago.

Garside, R. G. (1980) *The Architecture of Digital Computers*, Clarendon, Oxford.

Gehani, N. (1984) *Ada: Concurrent Programming*, Prentice-Hall, Englewood Cliffs, NJ.

Genrich, H. J. and Lautenbach, K. (1981) 'System Modelling with High-level Petri Nets', *Theoretical Computer Science* **13** 109–136.

Giloi, W. K., Balaci, R. and Behr, P. (1978) *APL\*DS*, Bericht 78–21 der Technischen Universität Berlin.

Goldberg, A. and Robson, D. (1983) *Smalltalk-80, the Language and its Implementation*, Addison-Wesley, Reading, Mass.

Golden, D. G. and Schoeffler, J. D. (1973) 'GSL — a Combined Continuous and Discrete Simulation Language', *Simulation* **20** 1 1–8.

Goldstine, H. H. (1972) *The Computer from Pascal to von Neumann*, Princeton University Press, Princeton, NJ.

Gonnet, G. H. (1976) 'Heaps Applied to Event Driven Mechanisms', *Comm. ACM* **19** 7 417–418.

Goodenough, J. B. (1975) 'Exception Handling: Issues and a Proposed Notation', *Comm. ACM* **18** 12 683–696.

Gordon, G. (1975) *The Application of GPSS V to Discrete System Simula-*

*tion*, Prentice-Hall, Englewood Cliffs, NJ.

Gordon, G. (1978) 'The Development of the General Purpose Simulation System (GPSS)', *ACM Sigplan Notices* **13** 8 183–198.

Greenberg, S. (1972) *GPSS Primer*, Wiley, New York.

Greenspan, D. (1973) *Discrete Models*, Addison-Wesley, Reading, MA.

Gribbin, J. (1984) *In Search of Schrödinger's Cat: Quantum Physics and Reality*, Bantam, Toronto.

Griswold, R. E. (1975) *String and List Processing in SNOBOL4*, Prentice-Hall, Englewood Cliffs, NJ.

Haggett, P. and Chorley, R. J. (1967) 'Models, Paradigms and the New Geography', Chapter 1 of Chorley, R. J. and Haggett, P. (Eds) (1967) *Models in Geography*, 19–41, Methuen, London.

Hammersley, J. M. and Handscomb, D. C. (1964) *Monte Carlo Methods*, Methuen, London.

Hamming, R. (1962) *Numerical Methods for Scientists and Engineers*, McGraw-Hill, New York.

Hartley, M. G. (1975) 'Introduction', in Hartley, M. G. (Ed.) (1975) *Digital Simulation Methods*, Peter Peregrinus, Stevenage, Hertfordshire, 1–22.

Heerman, D. W. (1986) *Computer Simulation Methods in Theoretical Physics*, Springer-Verlag, Berlin.

Heidorn, G. E. (1974) 'English as a Very High Level Language for Simulation Programming', *ACM Sigplan Notices* **9** April 91–100.

Helsgaun, K. (1980) 'DISCO — a SIMULA-based Language for Continuous, Combined and Discrete Simulation', *Simulation* **35** July, 1–12.

Hennessy, M. (1984) 'An Algebraic Theory of Fair Asynchronous Communicating Processes', University of Edinburgh, Dept. Computer Science, Internal Report CSR-171-84, October 1984.

Henriksen, J. O. (1977) 'An Improved Events List Algorithm', *Proc. Winter Simulation Conference*, Tampa, Florida, 547–557.

Henriksen, J. O. (1983) 'Event List Management — a Tutorial', *Proc. 1983 Winter Simulation Conference*, IEEE, 543–551.

Henriksen, J. O. and Crain, R. C. (1982) *GPSS/H User's Manual*, Second Edition, Wolverine Software Corporation, Annandale, VA.

Hill, D. D. and Coelho, D. R. (1987) *Multi-level Simulation for VLSI Design*, Kluwer, Boston.

Hills, P. R. (1967) 'Simon — A Computer Simulation Language in Algol', in Hollingdale, S. H. (1967) 105–115.

Hills, P. R. (1973) *An Introduction to Simulation using Simula*, Publication No. S 55, Norwegian Computing Center, Oslo.

Hills, P. R. and Birtwistle, G. M. (1975) *SIMON 75: a Simulation Language in Simula*, Robin Hills (Consultants), Camberley, Surrey.

Hoare, C. A. R. (1985) *Communicating Sequential Processes*, Prentice-Hall, Englewood Cliffs, NJ.

Hobbs, J. R. and Moore, R. (Eds) (1983) *Contributions in Artificial Intelligence* **1**, Ablex, Norwood, NJ.

Hockney, R. W. and Eastwood, J. W. (1981) *Computer Simulation using*

*Particles*, McGraw-Hill, New York.

Hodges, A. (1985) *Alan Turing: the Enigma of Intelligence*, Counterpoint, Unwin, London.

Hofstadter, D. R. (1980) *Gödel, Escher, Bach: an Eternal Golden Braid*, Penguin Books, Harmondsworth.

Hogeweg, P. (1978) 'Simulating the Growth of Cellular Forms', *Simulation* **31** September, 90–96.

Hogeweg, P. (1980) 'Locally Synchronised Developmental Systems: Conceptual Advantages of Discrete Event Simulation', *International Journal of General Systems* **6** 57–73.

Hogeweg, P. and Hesper, B. (1979) 'Heterarchical, Selfstructuring Simulation Systems: Concepts and Applications in Biology' in Zeigler, B. P., *et al.* (1979) 221–232.

Holbaek-Hanssen, E., Handlykken, P. and Nygaard, K. (1977) *System Description and the DELTA Project*, Report No. 4, Publication 523, Norwegian Computer Center, Oslo.

Hollingdale, S. H. (Ed.) (1967) *Digital Simulation in Operational Research*, NATO Conference, Hamburg, 1965.

Holt, R. C. (1983) *Concurrent Euclid, the Unix System and Tunis*, Addison-Wesley, Reading, Mass.

Hooper, J. W. (1986) 'Strategy-related Characteristics of Discrete-event Languages and Models' *Simulation* **46** 4 153–159.

Hooper, J. W. and Reilly, K. D. (1982) 'An Algorithmic Analysis of Simulation Strategies', *International Journal of Computer and Information Sciences* **11** 2 101–122.

Hooper, J. W. and Reilly, K. D. (1983) 'The GPSS–GASP Combined (GGC) System', *International Journal of Computer and Information Sciences* **12** 2 111–136.

Houle, P. A. and Franta, W. R. (1975) 'On the Structural Concepts of Simula', *Australian Computer Journal* **7** 1 39–45.

Hull, T. E. and Dobell, A. R. (1962) 'Random Number Generation', *SIAM Review* **4** 3 230–254.

Hurrion, R. D. (Ed.) (1986) *Simulation: Applications in Manufacturing*, IFS (Publications), Bedford, UK.

Hurst, N. R. and Pritsker, A. A. B. (1973) 'Simulation of a Chemical Reaction Process Using GASP IV', *Simulation* **21** September 71–75.

Hutchinson, G. K. (1975) 'Introduction to the Use of Activity Cycles as a Basis for System's Decomposition and Simulation', *ACM Simuletter* **7** 1 15–20.

Ichbiah, J. D. and Morse, S. P. (1972) 'General Concepts of the Simula 67 Programming Language', *Annual Review in Automatic Programming* **7** 1 65–93.

Ichbiah, J. D., Barnes, J. G. P., Haliard, J. C., Krieg-Brueckner, B., Roubine, O. and Wichmann, B. A. (1979) 'Rationale for the Design of the ADA Programming Language', *ACM Sigplan Notices*, **14** 6 June Part B.

Ignall, E. and Kolesar, P. (1979) 'On Using Simulation to Extend OR/MS

Theory: the Symbiosis of Simulation and Analysis', in Adam and Dogramaci (1979), Chapter 15, 223–234.

Jefferson, D. and Sowizral, H. (1985) 'Fast Concurrent Simulation Using the Time Warp Mechanism', in *Proc. SCS Conference on Distributed Simulation*, San Diego, CA, 63–69.

Jensen, K. (1981) 'Coloured Petri Nets and the Invariant-method', *Theoretical Computer Science* **14** 317–336.

Johnson-Laird, P. N. and Wason, P. C. (Eds) (1977) *Thinking: Readings in Cognitive Science*, Cambridge University Press.

Jonassen, A. and Dahl, O.-J. (1975) 'Analysis of an Algorithm for Priority Queue Administration', *BIT* **15** 409–422.

Jones, D. W. (1986) 'An Empirical Comparison of Priority-queue and Event-set Implementations', *Comm. ACM* **29** 4 300–311.

Jones, D. W. (1987) 'A Note on Bottom-up Skew Heaps', *SIAM Journal of Computing*, **16** 1 108–110.

Kahn, G. and MacQueen, D. B. (1977) 'Coroutines and Networks of Parallel Processes', in *Proceedings IFIP 77*, Gilchrist, B. (Ed.), North Holland, Amsterdam, 993–998.

Kaubisch, W.-H., Perrott, R. H. and Hoare, C. A. R. (1976) 'Quasi-parallel Programming', *Software — Practice and Experience* **6** 341–356.

Kelley, D. H. and Buxton, J. N. (1962) 'Montecode — an Interpretive Program for Monte Carlo Simulations', *Computer Journal* **5** 88–93.

Kempf, K. G. (1983) 'Chess: AI :: Snooker : SB', *AISB Quarterly*, No. 46, Winter 82/83, 17–20.

Kindler, E. (1981) 'A Formalization of some Simulation Language Concepts', *International Journal of General Systems* **6** 183–190.

Kingston, J. H. (1984) *Analysis of Algorithms for the Simulation Event List*, Ph.D. Thesis, Basser Dept. of Computer Science, University of Sydney.

Kingston, J. H. (1985) 'Analysis of Tree Algorithms for the Simulation Event List', *Acta Informatica* **22** 1 15–33.

Kingston, J. H. (1986) 'The Amortized Complexity of Henriksen's Algorithm', *BIT* **26** 156–163.

Kiviat, P. J., Villanueva, R. and Markowitz, H. M. (1973) *SIMSCRIPT II.5 Programming Language*, CACI Inc., Los Angeles.

Kleijnen, J. P. C. (1978) Communication to the Editor, *Management Science* **24** 1772–1774.

Kleijnen, J. P. C. (1987) *Statistical Tools for Simulation Practitioners*, Marcel Dekker, New York.

Kleinrock, L. (1975) *Queueing Systems*, **1**, 'Theory', Wiley, New York.

Knuth, D. E. (1973a) *The Art of Computer Programming*, **1**, 'Fundamental Algorithms', Second Edition, Addison-Wesley, Reading, Mass.

Knuth, D. E. (1973b) *The Art of Computer Programming*, **3**, 'Sorting and Searching', Addison-Wesley, Reading, Mass.

Knuth, D. E. (1974) 'Structured Programming with **go to** Statements', *ACM Computer Surveys* **6** 4 261–301.

Knuth, D. E. (1977) 'Deletions That Preserve Randomness', *IEEE Trans on Software Engineering* **SE-3** 5 351–359.

Knuth, D. E. (1981) *The Art of Computer Programming*, **2**, 'Seminumerical Algorithms', Second Edition, Addison-Wesley, Reading, Mass.

Knuth, D. E. and McNeley, J. L. (1964a) 'SOL — a Symbolic Language for General-purpose Systems Simulation', *IEEE Trans on Computers* **C-13** 401–408.

Knuth, D. E. and McNeley, J. L. (1964b) 'A Formal Definition of SOL', *IEEE Trans on Computers* **C-13** 409–414.

Kolmogorov, A. N. (1968a) 'Three Approaches to the Quantitative Definition of Information', *International Journal of Computer Mathematics* **2** 157–168. A translation from *Problemy Peredachi Informatsii* **1** (1965) 1 3–11.

Kolmogorov, A. N. (1968b) 'Logical Basis for Information Theory and Probability Theory', *IEEE Trans. on Information Theory* **IT-14** 5 662–664.

Krasnow, H. S. and Merikallio, R. A. (1964) 'The Past, Present, and Future of General Simulation Languages', *Management Science*, **11** 2 236–267.

Kreutzer, W. (1986) *System Simulation: Programming Styles and Languages*, Addison-Wesley, Sydney.

Kuck, D. J. (1978) *The Structure of Computers and Computations*, Wiley, New York.

Kuhn, T. S. (1970) *The Structure of Scientific Revolutions*, Second Edition, University of Chicago Press.

LaLonde, W. R., Pugh, J. R. and Thomas, D. A. (1985) *Smalltalk: Discovering the System*, Technical Report SCS-TR-80, Carleton University, Ottawa.

Lamprecht, G. (1983) *Introduction to Simula 67*, Vieweg, Braunschweig.

Lapalme, G. and Vaucher, J. (1981) 'Une Implantation Efficace de l'Ordonnancement Conditionnel', *RAIRO Informatique* **15** 3 255–285.

Laski, J. G. (1965) 'On Time Structure in (Monte Carlo) Simulations', *Operational Research Quarterly* **16** 3 329–339.

Laski, J. G. (1968) 'Two Proposals Towards Discrete Modelling', in Buxton, J. N. (Ed.) (1968), 175–197.

Lauer, P. E. and Shields, M. W. (1978) 'Abstract Specification of Resource Accessing Disciplines, Adequacy, Starvation, Priority and Interrupts', *ACM Sigplan Notices* **13** 12 (December) 41–59.

Laurini, R. (1979) 'Introduction au Multipilotage Urbain', *RAIRO Automatique*, **13** 3 303–322.

Laurini, R. (1981) 'A Primer of Predicate-driven Simulation', private communication.

Law, A. M. and Kelton, W. D. (1982) *Simulation Modeling and Analysis*, McGraw-Hill, New York.

Lenz, J. E. (Ed.) (1986) *Simulation in Manufacturing*, Proc. 2nd International Conference, Chicago, June 1986.

Levinson, M. R. (1973) 'Simulation with ALGOL 68', *Algol Bulletin* **36**.4.2 25–27 November.

Levinson, M. R. (1974) 'Simulation with ALGOL 68', *Algol Bulletin* **38**.4.2

43–44 December.

Lewandowski, A. (1982) 'Issues in Model Validation', *Angewandte Systemanalyse*, **3** 2–11.

Lindenmayer, A. (1968) 'Mathematical Models for Cellular Interactions in Development', *Journal of Theoretical Biology* **18** 280–315.

Lindsey, C. H. (1974a) 'Partial Parameterization', *Algol Bulletin* **37** 4.2 24–26.

Lindsey, C. H. (1974b) 'Modals', *Algo Bulletin* **37** 4.3 26–29.

Lindsey, C. H. and van der Meulen, S. G. (1977) *Informal Introduction to ALGOL 68*, Revised Edition, North Holland, London.

Mäkinen, E. (1986) *Splay Trees as Priority Queues*, Report A168, Department of Mathematical Sciences, University of Tampere, Finland.

Mandelbrot, B. B. (1977) *Fractals: Form, Chance and Dimension*, Freeman, San Francisco.

Markowitz, H. M., Hausner, B. and Karr, H. W. (1963) *SIMSCRIPT: a Simulation Programming Language*, Prentice-Hall, Englewood Cliffs, NJ.

Marlin, C. D. (1980) *Coroutines*, Lecture Notes in Computer Science, **95**, Springer-Verlag, Berlin.

Marsaglia, G. (1968) 'Random Numbers Fall Mainly in the Planes', *Proc. Nat. Acad. Sci.* **61** 25–28.

Marsaglia, G. (1985) 'A Current View of Random Number Generators', In Billard, L. (Ed.) (1985) *Computer Science and Statistics: Proc. 16th Symposium on the Interface*, Atlanta, Ga., March 1984, 3–10.

Maryanski, F. (1980) *Digital Computer Simulation*, Hayden, Rochelle Park, NJ.

Mathewson, S. C. (1975) 'Simulation Program Generators', *Simulation* **23** 6 181–189.

Mathewson, S. C. (1985) 'Simulation Program Generators: Code and Animation on a P.C.', *Journal of the Operational Research Society* **36** 7 583–589.

Maynard Smith, J. (1974) *Models in Ecology*, Cambridge University Press.

Mendelbaum, S. C. and Madaule, F. (1976) 'Automata as Structured Tools for Real Time Programming', in *Proceedings of 1975 IFAC/IFIP Workshop on Real Time Programming*, Boston, Mass., 59–65.

Metropolis, N., Howlett, J. and Rota, G.-C. (Eds) (1980) *A History of Computing in the Twentieth Century*, Academic Press, New York.

Metropolis, N. and Ulam, S. (1949) 'The Monte Carlo Method', *J. Amer. Statist. Assoc.* **44** 335.

van der Meulen, S. G. (1977) 'Algol 68 Might-have-beens', Proc. Algol 68 Conference, Strathclyde, *ACM Sigplan Notices* **12** 6 1–18.

Minsky, M. (1968) 'Matter, Mind and Models', in Minsky, M. (Ed.) *Semantic Information Processing*, MIT Press, Cambridge, Mass. 425–432.

Mitrani, I. (1982) *Simulation Techniques for Discrete Event Systems*, Cambridge University Press.

Monod, J. (1972) *Chance and Necessity*, William Collins, Glasgow.

Moody, K. and Richards, M. (1980) 'A Coroutine Mechanism for BCPL', *Software — Practice and Experience* **10** 765–771.

Moser, J. G. (1986) 'Integration of Artificial Intelligence and Simulation in a Comprehensive Decision-support System', *Simulation* **47** 6 223–229.

Moss, J. E. B. (1985) *Nested Transactions*, MIT Press, Cambridge, MA.

Musielak, H. und Stoessel, M. (1979) 'Vergleich von Simulationssprachen', *Elektronische Rechenanlagen*, **21** 1 23–28.

Myhrhaug, B. (1965) *Sequencing Set Efficiency*, Pub. A9, Norwegian Computing Center, Forskningsveien 1B, Oslo 3.

McCormack, W. M. (1979) *Analysis of Future Event Set Algorithms for Discrete Event Simulation*, Ph.D. Dissertation, Syracuse University.

McCormack, W. M. and Sargent, R. G. (1979) 'Comparison of Future Event Set Algorithms for Simulations of Closed Queueing Systems', in Adam, N. R. and Dogramaci, A. (Eds) (1979).

McCormack, W. M. and Sargent, R. G. (1981) 'Analysis of Future Event Algorithms for Discrete Event Simulation', *Comm. ACM* **24** 12 801–812.

McDermott, D. V. (1982) 'A Temporal Logic for Reasoning About Plans and Actions', *Cognitive Science* **6** 101–155.

McDermott, D. V. (1983) 'Reasoning about Plans', in Hobbs and Moore (1983) 269–317.

MacDougall, M. H. (1975) 'System Level Simulation', in Breuer, M. A. (Ed.) *Languages, Simulation and Data Base*, Pitman, San Francisco, 1–116.

MacDougall, M. H. and McAlpine, J. S. (1973) 'Computer System Simulation with Aspol', in *Proceedings of the Symposium on the Simulation of Computer Systems*, June 19–20, 1973, 92–103.

McGettrick, A. D. (1978) *ALGOL 68: a First and Second Course*, Cambridge University Press, Cambridge.

MacKay, D. M. (1969) *Information, Mechanism and Meaning*, MIT, Cambridge, MA.

MacLaren, M. D. and Marsaglia, G. (1965) 'Uniform Random Number Generators', *Journal ACM* **12** 83–89.

McQuarrie, D. A. (1967) *Stochastic Approach to Chemical Kinetics*, Methuen, London.

Nance, R. E. (1971) 'On Time Flow Mechanisms for Discrete Event Simulation', *Management Science* **18** 1 59–73.

Nance, R. E. (1979) 'Model Representation in Discrete Event Simulation: Prospects for Developing Documentation Standards', in Adams and Dogamaci (1979) 83–97.

Nance, R. E. (1980) *The Time and State Relationships in Simulation Modeling*, Combat Systems Department, Naval Surface Weapons Center, Dahlgren, VA.

Nance, R. E. (1981) 'The Time and State Relationships in Simulation Modeling', *Comm. ACM* **24** 4 173–179.

Nance, R. E. and Overstreet, C. M. (1986) 'Diagnostic Assistance using Digraph Representations of Discrete Event Simulation Model Specifications' SRC-86-001, Systems Research Center, Department of Computer Science, Virginia Tech., Blacksburg, VA.

Nevalainen, O. and Teuhola, J. (1978) 'The Efficiency of Two Indexed Priority Queue Algorithms', *BIT* **18** 3 320–333.

Nevalainen, O. and Teuhola, J. (1979) 'Priority Queue Administration by Sublist Index', *Computer Journal* **22** 3 220–225.

Nicholls, J. E. (1975) *The Structure and Design of Programming Languages*, Addison-Wesley, Reading, Mass.

Niederreichholz, J. und Stockheim, F. (1979) 'Kostenvergleich der Simulationssprachen GPSS 1100 und SIMULA 1', *Angewandte Informatik*, Teil 1, Januar, 1–8.

Nielsen, M., Plotkin, G. and Winskel, G. (1981) 'Petri Nets, Event Structures and Domains, Part I', *Theoretical Computer Science* **13** 85–108.

Nygaard, K. (1986) 'Basic Concepts in Object Oriented Programming', *ACM Sigplan Notices* **21** 10 (October) 128–132.

Nygaard, K. and Dahl, O.-J. (1978) 'The Development of the SIMULA Languages', *ACM Sigplan Notices* **13** 8 245–272.

O'Keefe, R. M. (1985) 'Comment on "Complexity Analyses of Event Set Algorithms"', *Computer Journal* **28** 5 496-497.

Ören, T. I. (1977) 'Software for Simulation of Combined, Continuous and Discrete Systems', *Simulation* **28** 2 33–45.

Ören, T. I. and Zeigler, B. P. (1979) 'Concepts for Advanced Simulation Methodologies', *Simulation* **32** March 69–82.

Organick, E. I. (1973) *Computer System Organisation: the Burroughs 5700/6700 Series*, Academic Press, New York.

Overstreet, C. M. and Nance, R. E. (1985) 'A Specification Language to Assist in Analysis of Discrete Event Simulation Models', *Comm. ACM* **28** 2 190–201.

Parnas, D. L. (1969) 'On Simulating Networks of Parallel Processes in which Simultaneous Events May Occur', *Comm. ACM* **12** 9 519–531.

Park, D. (1980) 'On the Semantics of Fair Parallelism', in *Abstract Software Specifications*, LNCS **86** 504–526 Springer Verlag, Berlin.

Paul, R. J. and Doukidis, G. I. (1986) 'Further Developments in the use of Artificial Intelligence Techniques which Formulate Simulation Problems', *Journal of the Operational Research Society* **37** 8 787–810.

Perrott, R. H., Raja, A. R. and O'Kane, P. C. (1980) 'A Simulation Experiment Using Two Languages', *Computer Journal* **23** 2 142–145.

Peterson, J. L. (1977) 'Petri Nets', *ACM Computing Surveys* **9** 3 223–252.

Peterson, J. L. (1980) 'A Note on Colored Petri Nets', *Information Processing Letters* **11** 1 40–43.

Petri, C. A. (1962) *Kommunikation mit Automaten*, Schriften des Institut fuer Instrumentelle Mathematik, Bonn.

Petrone, L. (1968) 'On a Simulation Language Completely Defined onto the Programming Language PL/I', in Buxton, J. N. (Ed.) (1968) 61–85.

Pidd, M. (1984) *Computer Simulation in Management Science*, Wiley, Chichester.

Poole, T. G. and Szymankiewicz, J. Z. (1977) *Using Simulation to Solve Problems*, McGraw-Hill, London.

Poston, T. and Stewart, I. (1978) *Catastrophe Theory and its Applications*, Pitman, London.

Preston, K., Jr. and Duff, M. J. B. (1984) *Modern Cellular Automata: Theory and Applications*, Plenum Press, New York.

Pritsker, A. A. B. (1974) *The GASP IV Simulation Language*, Wiley-Interscience, New York.

Pritsker, A. A. B. (1979) 'GASP: Present Status and Future Prospects', in Adam and Dogramaci (1979) 61–70.

Pritsker, A. A. B. and Pegden, C. D. (1979) *Introduction to Simulation and SLAM*, Wiley, New York.

Reddy, Y. V. R., Fox, M. S. and Husain, N. (1986) 'The Knowledge-based Simulation System', *IEEE Software* March 26–37.

Reeves, C. M. (1984) 'Complexity Analyses of Event Set Algorithms', *Computer Journal* **27** 1 72–79.

Reisig, W. (1984) 'What Operational Semantics is Adequate for Nonsequential Systems?' Digital Systems Laboratory, Research Report No. 30, Helsinki University of Technology.

Reisig, W. (1985) *Petri Nets: an Introduction*, Springer-Verlag, Berlin.

Reps, T. W. (1984) *Generating Language-based Environments*, MIT, Cambridge, MA.

Riddle, W. E. and Sayler, J. H. (1979) 'Modelling and Simulation in the Design of Complex Software Systems', in Zeigler, B. P., *et al.* (1979) 359–386.

Rivett, P. (1980) *Model Building for Decision Analysis*, Wiley, Chichester.

Rivett, B. H. P. (1983) 'Professor K. D. Tocher: a Personal Appreciation', *Journal of the Operational Research Society* **34** 4 265–270.

Rosenblueth, A. and Wiener, N. (1945) 'The Role of Models in Science', *Philosophy of Science* **12** 316–321.

Russell, E. C. (1983) *Building Simulation Models with SIMSCRIPT II.5*, CACI Inc., Los Angeles.

Sacerdoti, E. D. (1977) *A Structure for Plans and Behavior*, Elsevier North-Holland, New York.

Sampson, J. R. and Dubreuil, M. (1979) 'Design of Interactive Simulation Systems for Biological Modelling', in Zeigler, B. P., *et al.* (1979), 233–248.

Sayre, K. M. and Crosson, F. J. (Eds) (1963) *The Modeling of Mind: Computers and Intelligence*, Simon and Schuster, New York.

Schildt, H. (1986) *Advanced C*, Osborne McGraw-Hill, Berkeley, CA.

Schmidt, B. (1978) *GPSS–FORTRAN Version II*, Springer–Verlag, Heidelberg.

Schmidt, D. A. (1986) *Denotational Semantics: A Methodology for Language Development*, Allyn and Bacon, Boston, MA.

Schoffeniels, E. (1976) *Anti-Chance*, Pergamon, Oxford.

Schriber, T. J. (1974) *Simulation Using GPSS*, Wiley, New York.

Schruben, L. (1983) 'Simulation Modeling with Event Graphs', *Comm.*

*ACM* **26** 11 957–963.

Schwartz, J. T., Dewar, R. B. K., Dubinsky, E. and Schonberg, E. (1986) *Programming with Sets: an Introduction to SETL*, Springer-Verlag, New York.

Shannon, R. E. (1975) *Systems Simulation: the Art and Science*, Prentice-Hall, Englewood Cliffs, NJ.

Shearn, D. C. S. (1975) 'Discrete Event Simulation in Algol 68' *Software — Practice and Experience* **5** 279–293.

Sim, R. (1975) *CADSIM Users' and Reference Manual*, Imperial College Publications, London.

Simon, H. A. (1973) 'The Organisation of Complex Systems', in Pattee, H. H. (Ed.) (1973) *Hierarchy Theory: the Challenge of Complex Systems*, Braziller, New York, 1–28.

Sleator, D. D. and Tarjan, R. E. (1983) 'Self-adjusting Binary Trees', *Proc. ACM SIGACT Symposium on Theory of Computing*, ACM, New York, 235–245.

Sleator, D. D. and Tarjan, R. E. (1985) 'Self-adjusting Binary Search Trees', *Journal of the ACM*, **32** 3 652–686.

Sleator, D. D. and Tarjan, R. E. (1986) 'Self-adjusting Heaps', *SIAM Journal of Computing*, **15** 1 52–69.

Spriet, J. A. and Vansteenkiste, G. C. (1982) *Computer-aided Modelling and Simulation*, Academic Press, London.

Stasko, J. T. and Scott Vitter, J. (1987) 'Pairing Heaps: Experiments and Analysis', *Comm. ACM* **30** 3 234–249.

Stoy, J. (1977) *Denotational Semantics: the Scott–Strachey Approach to Programming Language Theory*, MIT Press, Cambridge, MA.

Stubbs, D. F. and Webre, N. W. (1987) *Data Structures with Abstract Data Types and Modula-2*, Brookes/Cole, Monterey, CA.

Tarjan, R. E. (1983) *Data Structures and Network Algorithms*, SIAM Philadelphia, PA.

Tarjan, R. E. (1985) 'Amortized Computational Complexity', *SIAM Journal for Algebraic and Discrete Methods* **6** 2 306–318.

Tarjan, R. E. (1987) 'Algorithm Design', *Comm. ACM* **30** 3 204–212.

Tausworthe, R. C. (1965) 'Random Numbers Generated by Linear Recurrence Modulo Two', *Mathematics of Computing* **19** 201–209.

Teichroew, D. and Lubin, J. F. (1966) 'Computer Simulation — Discussion of the Technique and Comparison of Languages', *Comm. ACM* **9** 10 723–741.

Tellaeche Bosch, R. and Downes, V. A. (1983) *Ada Simulation Library: User's Guide*, Research Report DOC 83/24, Department of Computing, Imperial College London.

Thom, R. (1975) *Structural Stability and Morphogenesis*, Benjamin, Reading, MA.

Tocher, K. D. (1963) *The Art of Simulation*, English Universities Press, London.

Tocher, K. D. (1964) 'Some Techniques of Model Building', in *Proceedings of IBM Scientific Computing Symposium on Simulation Models and*

*Gaming*, New York, 119–155.

Tocher, K. D. (1965) 'Review of Simulation Languages', *Operational Research Quarterly* **16** 2 189–217.

Tocher, K. D. (1969) 'Simulation: Languages', in *Progress in Operations Research*, Volume III, Aronofsky, J. S. (Ed.), 71–113, Wiley, New York.

Tocher, K. D. (1979) 'Keynote Address' *Proceedings of the 1979 Winter Simulation Conference*, IEEE, New York, NY.

Tocher, K. D. and Owen, D. G. (1960) 'The Automatic Programming of Simulations', *Proc. IFORS Conference*, Aix-en-Provence, 50–67.

Törn, A. A. (1981) 'Simulation Graphs: a General Tool for Modeling Simulation Designs', *Simulation* **37** 6 187–194.

Törn, A. A. (1985) 'Simulation Nets, a Simulation Modeling and Validation Tool', *Simulation* **45** 2 71–75.

Treleaven, P. C., Brownbridge, D. R. and Hopkins, R. P. (1982) 'Data-driven and Demand-driven Computer Architecture', *ACM Computing Surveys* **14** 1 93–143.

Tsang, E. P. K. (1986a) *The Interval Structure of Allen's Time Logic*, Report CSCM-24, Cognitive Studies Centre, University of Essex.

Tsang, E. P. K. (1986b) *TLP — a Temporal Planner*, Report CSCM-27, Cognitive Studies Centre, University of Essex.

Turing, A. M. (1950) 'Computing Machinery and Intelligence', *Mind* **59** 433–460.

Turing, A. M. (1952) 'A Chemical Basis of Morphogenesis', *Phil. Trans. Roy. Soc.* **B237** 37–72.

Turner, C. (1986) 'Simulation and Execution of Planning Systems', *AISB Quarterly* No. 58, Summer 8–11.

Turner, R. (1984) *Logics for Artificial Intelligence*, Ellis Horwood, Chichester.

Ulam, S. M. (1980) 'von Neumann: the Interaction of Mathematics and Computing', in Metropolis, N. *et al.* (Eds) (1980) 93–99.

Unger, B. W., Lomow, G. A. and Birtwistle, G. M. (1984) *Simulation Software and Ada*, Society for Computer Simulation, La Jolla, CA.

Varela, F. J. and Goguen, J. A. (1978) 'The Arithmetic of Closure', in *Progress in Cybernetics and Systems Research*, Volume III, Trappl, R., (Ed.), 48–64, Hemisphere Publications, Washington D.C.

Vaucher, J. G. (1971) 'Simulation Data Structures using Simula 67', *Proc. Winter Simulation Conference*, New York, 255–260.

Vaucher, J. G. (1973) 'A "Wait Until" Algorithm for General Purpose Simulation Languages', *Proc. Winter Computer Simulation Conference*, 77–83.

Vaucher, J. G. (1977) 'On the Distribution of Event Times for the Notices in a Simulation Event List', *INFOR* **15** 2 171–182.

Vaucher, J. G. (1985): 'Views of Modelling: Comparing the Simulation and AI Approaches', *Proc. SCS Conference on AI, Graphics and Simulation*, San Diego, CA, 3–7.

Vaucher, J. G. and Duval, P. (1975) 'A Comparison of Simulation Event

List Algorithms', *Comm. ACM* **18** 4 223–230.

Vessey, I. and Weber, R. (1986) 'Structured Tools and Conditional Logic: an Empirical Investigation', *Comm. ACM* **29** 1 48–57.

Virjo, A. (1972) 'A Comparative Study of Some Discrete-event Simulation Languages', Proc. Norddata 72 Conference, Helsinki, pub. Norwegian Computing Center, 1532–1564.

Warren, H. J., Low, R. A., Hughes, P. D. and Gray, G. J. (1985) 'Autosim: an Automatic Simulation Program Generator', *Mathematics and Computers in Simulation* **27** 107–114.

Wells, M. B. (1980) 'Reflections on the Evolution of Algorithmic Language', in Metropolis, N. *et al.* (Eds) (1980) 275–287.

Wells, D. (1986) *The Penguin Dictionary of Curious and Interesting Numbers*, Penguin, Harmondsworth.

Wiener, N. (1948) *Cybernetics: or Control and Communication in the Animal and the Machine*, MIT Press, Cambridge, MA.

van Wijngaarden, A., Mailloux, B. J., Peck, J. E. L., Koster, C. H. A., Sintzoff, M., Lindsey, C. H., Meertens, L. G. L. T. and Fisker, R. G. (Eds) (1976) *Revised Report on the Algorithmic Language Algol 68*, Springer-Verlag, Berlin.

Williams, J. W. J. (1963) 'E. S. P. The Elliott Simulator Package', *Computer Journal* **6** 328–331.

Williams, J. W. J. (1964) 'Algorithm 232', *Comm. ACM* **7** 6 347–348.

Winskel, G. (1980) 'Events in Computation', Ph.D. thesis, University of Edinburgh, Department of Computer Science, CST-10-80, December 1980.

Wirth, N. (1976) *Algorithms + Data Structures = Programs*, Prentice-Hall, Englewood Cliffs, NJ.

Wirth, N. (1982) *Programming in Modula-2*, Springer-Verlag, Berlin.

Wolf, A. (1983) 'Simplicity and Universality in the Transition to Chaos', *Nature* **305** 15 182–183.

Wyman, F. P. (1975) 'Improved Event-scanning Mechanisms for Discrete Event Simulation', *Comm. ACM* **18** 6 350–353.

Yakowitz, S. J. (1977) *Computational Probability and Simulation*, Addison-Wesley, Reading, Mass.

Young, S. J. (1982) *Real Time Languages: Design and Development*, Ellis Horwood, Chichester.

Zeigler, B. P. (1976) *Theory of Modelling and Simulation*, Wiley, New York.

Zeigler, B. P. (1984a) 'Multifaceted Modeling Methodology: Grappling with the Irreducible Complexity of Systems', *Behavioral Science* **29** 169–178.

Zeigler, B. P. (1984b) *Multifaceted Modeling and Discrete Event Simulation*, Academic Press, New York.

Zeigler, B. P. (1985) 'Discrete Event Formalism for Model Based Distributed Simulation', in *Proc. SCS Conference on Distributed Simulation*, San Diego, CA, 3–7.

Zeigler, B. P., Elsas, M. S., Klir, G. J. and Ören, T. I. (Eds) (1979)

*Methodology in Systems Modelling and Simulation*, North Holland, Amsterdam.

Zisman, M. D. (1978) 'Use of Production Systems for Modeling Asynchronous, Concurrent Processes', in Waterman, D. A. and Hayes-Roth, F. (Eds) (1978) *Pattern-directed Inference Systems*, Academic Press, New York, 53–68.

# Index